Renewing the Promise:

Research-Intensive Universities and the Nation

A report prepared by the
President's Council of Advisors on Science and Technology

December 1992

The President's Council of Advisors on Science and Technology

Members

D. ALLAN BROMLEY
Assistant to the President for Science and Technology, and
Director of the Office of Science and Technology Policy
Executive Office of the President *(Chairman)*

HAROLD SHAPIRO
President, Princeton University *(Vice Chairman)*

NORMAN BORLAUG
Distinguished Professor, Department of Soils and Crop Sciences
Texas A&M University

SOLOMON BUCHSBAUM
Senior Vice President, Technology Systems, AT&T Bell Laboratories

CHARLES DRAKE
Albert Bradley Professor of Earth Sciences and Professor of Geology
Dartmouth College

RALPH GOMORY
President, The Sloan Foundation

MARY L. GOOD
Senior Vice President, Research and Technology, Allied-Signal, Inc.

PETER LIKINS
President, Lehigh University

THOMAS LOVEJOY
Assistant Secretary for External Affairs, The Smithsonian Institution

THOMAS J. MURRIN
Dean, School of Business and Administration, Duquesne University

JOHN McTAGUE
Vice President, Technical Affairs, Ford Motor Company

DANIEL NATHANS
Professor of Molecular Biology and Genetics
Johns Hopkins University School of Medicine

DAVID PACKARD
Chairman of the Board, Hewlett-Packard Company

Emeritus Members

BERNADINE HEALY
Director, National Institutes of Health

WALTER MASSEY
Director, National Science Foundation

Staff

VICKIE V. SUTTON
Assistant Director for Management and the Science Councils
Office of Science and Technology Policy

ALICIA K. DUSTIRA
Executive Secretary
President's Council of Advisors on Science and Technology
Office of Science and Technology Policy

THE WHITE HOUSE

WASHINGTON

November 23, 1992

Dear Mr. President:

I have the pleasure of transmitting to you, on behalf of the President's Council of Advisors on Science and Technology, our latest report, *Renewing the Promise: Research-Intensive Universities and the Nation*. For this study, we considered research-intensive universities as the approximately one hundred and fifty institutions that conduct a significant share of our country's basic research and that produce our country's leading scientific and engineering talent. This makes them the foundation of the nation's science and technology enterprise.

The "promise" refers to the vision contained in Dr. Vannevar Bush's 1945 report to President Truman: that our leading universities, with strong public support, could generate the intellectual and human capital that would, in turn, enable the United States to become a more prosperous, healthier and more secure Nation. We followed that vision and the extraordinary accomplishments of the United States over the past half-century -- in technology and industry, in medicine and health, in agriculture and management of our environment -- attest to the fulfillment of the promise.

In our rapidly changing world, the United States faces ever increasing economic and technological challenges and environmental, public health, and national security concerns that differ from those of the past. The Council believes that, more than ever, our Nation will depend on its colleges and universities for the generation of the new knowledge and talent needed to maintain world leadership. The "promise" must be renewed.

The Council is concerned, however, that the relationship between the Federal government and the research-intensive universities is under stress, and that the universities are not as well-positioned as they should be to help meet the Nation's future needs.

In its report, the Council identifies weak points in the system and suggests some corrective measures for universities and Federal agencies, as well as state governments, industry, and other supporters of universities. We believe that these recommendations, if followed, would go a long way toward re-strengthening the universities and enabling them, in partnership with the Federal government, to make the greatest possible contribution to the Nation.

The Council believes that the issues raised by the report are of fundamental, non-partisan importance. We hope, therefore, that you will not only find the report helpful to your own thinking about the long-term interests of the Nation, but that you will commend it to the President-elect for his consideration. Furthermore, if you agree, the Council will release the report to the public. It should stimulate a healthy national debate on the issues.

Although the Council concluded that the U.S. research and development (R&D) system is basically sound and continues to serve us well, we feel that there is also a need to look beyond the issues addressed in this report. Just as the end of World War II was an opportune time for Vannevar Bush and his colleagues to looked ahead at the role of R&D in our national life, the end of the Cold War warrants a new examination of the Nation's R&D system: its rationale, goals, organization, funding and administrative mechanisms.

We believe that the Council should begin to undertake such a re-examination as soon as possible. Here, too, we hope that you would urge the President-elect to instruct his Council to follow such a course in the interest of the Nation.

This report views the relationships between the Federal government and the research-intensive universities from the outside. I am transmitting to you, in parallel with this report, a companion report prepared by an interagency working group of the Federal Coordinating Council for Science, Engineering and Technology (FCCSET) that views these relationships from inside the Federal government. Together, they represent a comprehensive review of this important area.

Sincerely yours,

D. Allan Bromley
The Assistant to the President
for Science and Technology
and Chair,
President's Council of Advisors on Science and Technology

The President
The White House
Washington, DC 20500

Enclosure

EXECUTIVE OFFICE OF THE PRESIDENT
OFFICE OF SCIENCE AND TECHNOLOGY POLICY
WASHINGTON, D.C. 20506

November 18, 1992

Dear Dr. Bromley:

We are pleased to transmit to you the report on Research-Intensive Universities, entitled *Renewing the Promise: Research-Intensive Universities and the Nation*, produced by the President's Council of Advisors on Science and Technology (PCAST).

The report focuses on a particular class of institutions, the research-intensive universities, because of their particular role in the nation's Research and Development (R&D) enterprise. Because of their important relationship with the Federal government, the health and future of those universities are inherently of significant concern in public policy matters. But our focus in no way implies that other types of institutions of higher education do not play critical roles in the larger educational system, particularly as recruiters and developers of the nation's talent.

The issues facing us are numerous and complex. We have sought to identify those that are the most pressing but that can be addressed in a realistic fashion, as well. Some of our findings and recommendations may not be popular, but a positive future sometimes requires painful treatment. At the very least, we hope that the report will stimulate and contribute to a healthy public debate. Given the critical nature of the present situation, we hope that our recommendations will lead to a productive readjustment of the current system.

We have also recommended a next step. Beyond the immediate issues, there is a need for a very fundamental look at the place of R&D in our nation's future and how it should be planned, supported, carried out, and managed. Such a recommendation should be high on PCAST's agenda at the outset of the new Administration.

Sincerely yours,

David Packard
Chairman, Committee on
Research-Intensive Universities

Harold T. Shapiro
Vice-Chairman, Committee on
Research-Intensive Universities

Enclosure

The Honorable D. Allan Bromley
Assistant to the President for
Science & Technology
Executive Office of the President
Washington, DC 20500

TABLE OF CONTENTS

Preface

Earlier this year, the President's Council of Advisors on Science and Technology (PCAST) turned its full attention to a core component of the nation's science and technology enterprise: those institutions of higher education that produce simultaneously the bulk of America's most highly trained scientists and engineers and a substantial fraction of our country's new scientific knowledge. Because most faculty at those institutions devote considerable efforts and resources to research as well as teaching, we refer to them as "research-intensive universities." The term applies to somewhat more than one hundred and fifty universities from among more than three thousand institutions of higher education. Their direct and indirect contribution to our national well-being — economic growth, international competitiveness, and new jobs — is immeasurable.

For more than four decades, the research and the graduate and post-doctoral education capabilities of those universities have rested in large part on support from the federal government. Because the research-intensive universities both benefit the public and are supported by it, their health and productivity are key matters of public policy.

As we look to the future, we are convinced that America cannot continue to meet the challenges of international economic competition, of national security in a post-Cold War world, and of threats to the environment and to public health without the continued contributions of the research-intensive universities.

As the present time, however, we are concerned that America's research-intensive universities are not as well prepared as they should be to assume the responsibilities of the future. We are also concerned that the American public no longer has the confidence in the research-intensive universities that will enable those institutions to bear those responsibilities. This is not a healthy situation for the nation.

Thus, we have sought to identify the weak points that threaten the otherwise strong web that binds together the federal government and the tax-paying public with the research-intensive universities into a major scientific research and higher education partnership. We have also considered the relationship between universities and industry — one that is also critical to the nation's future. In this report, we present our findings and recommendations for strengthening the overall enterprise.

Our approach to the subject consisted of three parts. First, the Council took the unusual step of constituting itself as a committee of the whole for the project. We felt from the outset that the study required the widest possible range of institutional and scientific perspectives and experience: industrial, governmental, and academic and from the physical, natural, life, and social sciences and engineering.

Second, we solicited — and received — a wide range of information and views from various interested parties. At six meetings around the country we heard from undergraduates, graduate students, post-doctoral students, and untenured faculty members as well as senior faculty and university administrators. We heard views representing a cross-section of universities — large and small, public and private. Just as important to our findings, we gathered perspectives from four-year and two-year institutions and from minority institutions, all of which play critical roles with respect to the research-intensive universities. Representatives from companies in the areas in which our public meetings were held rounded out the picture, as did representatives of educational and scientific associations. Written comments were also received from a wide variety of individuals and institutions responding thoughtfully to our invitation to provide input and advice.

This public input was invaluable to the study. We express our deep gratitude to the many who so graciously hosted the meetings, who participated in them, and who sent us their observations and ideas. Without these contributions, many of our findings would not have emerged so clearly, if at all.

Third, we benefitted greatly from data and perspectives provided by a group of federal agencies whose policies and programs affect research and education. Meeting as an Ad Hoc Working Group on Research-Intensive Universities and the Federal Government of the Federal Coordinating Council for Science, Engineering, and Technology (FCCSET), they conducted their own study in parallel with ours. The Working Group is chaired by Deputy Secretary of Education David Kearns, with National Institutes of Health Director Bernadine Healy and National Science Foundation Director Walter Massey serving as Vice-Chairs. We are greatly appreciative to the FCCSET group for sharing its views and for the data and information it has provided. Any interpretation of those data in this report is, of course, our own.

Finally, we would be remiss indeed were we not to recognize the vital contribution made to this report by many members of the Office of Science and Technology Policy and PCAST organizations. In particular, our thanks go to Pierre Perrolle, William Raub, and James McCullough for outstanding staff work throughout this activity. We would also thank Alicia Tenuta and Philip Bolus for major contributions to the structure of the entire study, the organization of the public meetings, and the production of the report.

Executive Summary

Introduction

Vannevar Bush's 1945 report, *Science — The Endless Frontier*, promised the nation the possibility of a prosperous, healthy, and secure future if the United States made a substantial commitment to basic research. Bush proposed specifically that the federal government provide funds to the nation's colleges and universities so that they could generate new knowledge and scientific and engineering talent. That vision led ultimately to the establishment of the largest and strongest scientific and higher education enterprise that the world has ever seen.

The human and intellectual capital generated by U.S. colleges and universities over the past four decades has been the basis for a vast array of accomplishments that have touched the lives of all Americans: jobs in new industries; sophisticated consumer products; technological breakthroughs affecting transportation, communications, and entertainment; an abundance of food; advances in medicine and public health; increased security against the threat of nuclear annihilation; and improved protection against natural disasters and environmental degradation. The American taxpayer's investment in university-based research has paid off handsomely. The 1945 promise has been more than fulfilled.

Those accomplishments notwithstanding, the relationship between the American public and institutions of higher learning is showing serious signs of stress. This partnership of over one hundred and fifty research-intensive universities and the federal government has grown to be a research and educational enterprise of enormous size, scope, and complexity. Despite their success, or perhaps in part because of the ever increasing expectations derived from that success, universities are losing public confidence. The many partners in the overall enterprise — students and parents, the federal and state governments, foundations and industry, and faculty and scientific communities — increasingly are expressing discontent with the enterprise's current state and direction.

The problems underlying that discontent must be addressed. In today's rapidly changing world, the United States is confronted with shifting national security needs, increasing economic and technological challenges from other countries, and new environmental and public health concerns. If anything, America's dependence on its colleges and universities both for creating new knowledge and for training new talent is greater now than at any time in our history.

In reviewing the health of the nation's research-intensive universities and their relationship with the federal government, PCAST has focused on six main areas:

- the implications of a limited resource environment;

- the fundamental importance of teaching activities at research-intensive universities;

- the erosion of public trust and confidence in our universities;

- federal investment in university-based research;

- interactions between universities and industry;

- and the identification and development of exceptional talent for science and technology.

Findings and Recommendations

I. A Matter of Limits: Adapting to a New Resource Environment

Advancing the frontiers of knowledge is not, as it once may have been, a matter of intellectual luxury. In an era of relentless global economic competition, it is a national imperative. The United States must continue, as a nation, to invest in fundamental research. The returns on such research, although unknowable in advance, have proved enormous in the past and are likely to be even greater in the future.

A realistic assessment of the next several decades indicates, however, that no matter how firm our national resolve may be to invest in the future, resources will not expand as rapidly as our intellectual capacity to pursue promising research opportunities. It is unreasonable to expect that the system of research-intensive universities will continue to grow as it did during periods in the 1960s and 1980s.

The cross-pressure of expanding opportunities and constrained resources poses a risk for the United States: spreading its resources too thinly across its array of highly trained investigators and research-intensive universities. Most of our research-intensive universities aspire to excel in all or most areas of scholarship and education. As worthy as those aspirations might seem, they are likely to be ill-advised. They cannot be fully realized in an era of significantly constrained resources.

The relentless quest for an ever broader range of activities will inevitably destroy the most important aspect of American higher education and research — quality. We cannot afford to allow our higher education to decline to the levels of mediocrity that now characterize much of our precollege education.

So that research-intensive universities may continue to function productively in an environment of limited resources, PCAST urges them to:

> adopt more highly selective strategies based on a realistic appraisal of the future availability of resources and a commitment to meet world-class standards in all programs that are undertaken.

Such strategies will require universities to:

- eliminate or downsize some departments and specialties rather than sustain less than world-class activities in every area of science and engineering;

- collaborate with other academic, industrial, and governmental institutions in sharing instructional and research facilities and programs;

- build facilities or programs only where there are strong long-term prospects of sustaining them; and

- develop permanent institutional mechanisms for strategic planning that will foster a balance between activities and resources and among teaching, research, and other missions that are commensurate with society's values and the needs of the universities.

PCAST also urges federal agencies and other supporters of research-intensive universities to:

- refrain from encouraging universities to embark on new research or education programs or the building of facilities when there is little or no long-term prospect of sustaining those programs; and

- refrain from developing or implementing research or education programs that would increase the net capacity of the system of research-intensive universities.

II. A Matter of Education: The Importance of Teaching at Our Universities

More than ever in its history, America needs its institutions of higher education to train scientists and engineers and educate citizens who can fill increasingly technical jobs. It also needs them to educate citizens of a democratic society who can understand the principles of technologically-oriented public policy issues. Furthermore, it is essential that colleges and universities continue to adapt to the increasingly diverse educational paths of many of our citizens, and keep pace with their desire and need for life-long learning. Finally, U.S. scientists and engineers must be capable of working in an international environment, in which foreign language skills are very important.

Many higher education institutions, including research-intensive universities, are turning away from their educational mission, particularly from undergraduate education. We believe that many of the complaints of parents and students concerning the quality of undergraduate education are well-founded. Universities should reemphasize teaching in all its aspects, both inside and outside the classroom. In doing so, many institutions will have to curtail some of their research activities. However, if institutions are selective in allocating their resources, the net output of leading-edge research by our nationwide array of research-intensive universities need not decrease.

PCAST recommends that universities strengthen their educational functions by:

- increasing direct senior faculty involvement in teaching at both the undergraduate and graduate levels, and in counseling of students;

- balancing the contributions of teaching and interaction with students with those of research and public service in evaluating and rewarding faculty;

- placing less reliance on graduate teaching assistants and ensuring that they understand their subjects, are better prepared with regard to teaching methods, are able to express themselves well in English, and, like all faculty, are able to provide a supportive environment for women and minority students;

- increasing the involvement of undergraduates in hands-on frontier research; and

- placing greater stress on educating scientists and engineers in key foreign languages.

Even as universities reemphasize teaching, they and their patrons working together must keep tuition and educational costs from rising faster than the income of the average American family.

The federal agencies should ensure that their programs encourage universities to reemphasize education rather than discourage them — even inadvertently — from taking the measures recommended above.

III. A Matter of Public Trust: Restoring Confidence in Our Universities

Public trust in universities is eroding. There is public concern about the rising cost of higher education and about whether the value of higher education is commensurate with the costs. There is also concern about a possible decline in quality, especially in teaching. Both the public and the university community have been shaken by a few, widely publicized cases of scientific misconduct, conflicts of interest, and misuse of federal research funds.

In addition, expectations of what universities can do for the nation are rising faster than are the resources available to meet those expectations. In a time of great change, there is imperfect consensus at best — both locally and at a national level — about where universities should place their emphases. Before issues of resources and priorities can be fully addressed, public trust and confidence in universities must be restored.

In our view:

> the university community and its patrons, including federal agencies, must act in ways designed to preserve the core values that underlie the scientific and educational enterprise — free and creative pursuit of ideas and synergism between research and teaching. Current "politically correct" approaches in some universities are attacking the very foundations of higher education.

In addition to reemphasizing education (see Section II), we urge colleges and universities to:

- establish effective measures to eliminate fabrication, falsification, and plagiarism in scholarly work, and to eliminate fraud and waste in the administration of that work.

PCAST cautions, however, that excessive efforts to anticipate and eliminate all potential problems can lead to bureaucratic strictures that undermine or stifle scientific creativity.

Thomas Murrin and Allan Bromley discussing issues at a PCAST public meeting.

IV. A Matter of Wise Investments: Federal Support of University-Based Research

Short-term constraints on research and development (R&D) resources as well as the likelihood that resources will be limited well into the future have fueled a healthy national debate over priority setting in all sectors. The debate has covered many issues and generated many options. In the context of this report, PCAST addresses only those issues most germane to the relationship between research-intensive universities and the federal government and only those issues directly pertinent to mathematics, the sciences (including the social sciences), and engineering.

PCAST urges, first of all, that the following principles guide the relationship:

- Our nation must continue to invest enough in basic research to sustain world class accomplishments in all major areas of science and technology. In those areas where U.S. activity does not define the frontiers, it must be sufficiently close to those frontiers that we can exploit discoveries without delay, wherever and whenever they are made;

- A healthy federal government-university partnership — particularly in basic research — must be maintained. It has served the nation well since World War II and we have every reason to believe that it will do so in the future;

- Federal agencies must, in their interactions with universities at all levels, recognize that they are investing in institutions that simultaneously generate new knowledge and new talent, and that they are not just procuring research results;

- The federal government must pay the full costs of the university-based research it supports. To expect cost-sharing (except where it strengthens the research-teaching linkage or in other special circumstances) defeats the investment objective and may shift costs to students and their families;

- Federal funds for research should be allocated through competition on a merit basis. Basic research proposals from federal laboratories should compete with those from universities for federal funds. The growing practice of Congressional earmarking of R&D projects and facilities without merit review must cease and must not be initiated or encouraged by universities; and

- Much more attention should be focused on developing criteria for federal and other investments that this nation accepts as appropriate to its long-term goals and aspirations. Criteria, by their nature, have much greater stability and longevity than do priorities, which are extremely sensitive to changes in the environment in which they are made.

In the body of the report, PCAST makes specific recommendations to federal agencies based on the above principles. They address: full federal reimbursement of all legitimate indirect costs; the growing obsolescence of the physical infrastructure for university-based research; improvements in the administration of federal research support to universities; and reexamination of the role of federal laboratories.

The recommendations regarding infrastructure include establishing a substantial federal program for the repair and renovation of university research facilities. The program would be nationally competitive and merit reviewed and would require 50-50 matching from non-federal funds. Support would be available only to institutions that pledged to forego any federal funds earmarked without merit review. The program would operate only for a catch-up period of a few years to enable universities to bring their facilities up to an appropriate level of modernity. Beyond that period, universities would be expected to keep their facilities current on the basis of federal indirect cost support and other resources.

V. A Matter of Two Cultures: Universities and Industry

Some of the cultural differences that have long surrounded industrial research and university research have had the unfortunate effect of unnecessarily inhibiting the most effective interaction between industry and universities. The notion that each sector had its own well-delineated and isolated role and that new knowledge would flow as rapidly as necessary and in one direction from the universities to industry is completely at odds with today's world.

Today the pressure of international competition has introduced a critical time dimension into the system. The issue is not simply how much new knowledge is being generated but also how fast it is being translated into economically and socially beneficial products and processes. This argues for a much greater flow of information and, especially, of people in both directions between universities and industry.

Despite recent gains in building linkages between U.S. universities and industry, there are still too many individuals in each sector who hold negative perspectives, attitudes, and stereotypes with respect to the other sector. The nation cannot afford to have this situation persist, and much more effort is required to overcome it. Even fundamental research that is not expected to yield short-term answers to industry's scientific problems can benefit from being informed by the technical concerns of industry. Conversely, U.S. industry should have the benefit of easy and immediate access to the new knowledge and new talent generated by universities. Exchange of personnel, at all levels, is the surest answer to these problems.

Accordingly, PCAST recommends that:

- universities and industry together, through a wide range of concerted actions, should exchange scientists and engineers at all levels — especially their very best — between the two sectors for substantial periods of time and repeatedly throughout their careers.

VI. A Matter of the Best Scientific Talent: Tapping the Nation's Talent Base

Most important scientific discoveries have been made by a small number of very gifted people who had been provided the opportunity and time to pursue their intellectual interests. Brilliant young people can be found throughout the population — within both genders and every race and ethnic group, in every economic situation, and in every region of the country. Stronger public policies must be designed to identify scientifically-gifted persons at an early age and to help them develop their talents to the fullest, no matter what their circumstances.

If the United States is to continue to lead the world in basic scientific discoveries and in their exploitation, we will need to identify the most talented young people at the earliest possible time, encouraging their interest in advanced education and science, and giving them a sense of purpose as they pursue their education and career paths.

Considering that potential science and engineering talent is well distributed throughout the population, in diverse economic circumstances, and throughout the country, PCAST recommends:

- that the federal government develop programs to award a substantial number of portable undergraduate scholarships and graduate fellowships in science and engineering in each Congressional district. These awards would be made on a nationally competitive basis, using non-political, merit criteria.

As with the traditional program of National Science Foundation graduate fellowships, only citizens and permanent residents would be eligible; awardees would be able to attend any U.S. institution of their own choosing; and reasonable allowances would be included to cover institutional costs. The undergraduate program would include both beginning students and some who have completed one or two years of undergraduate work.

PCAST notes that federal, state, and local tax policies should bolster, not undermine, the nation's investments — both public and private — in human capital. PCAST recommends that:

- federal, state, and local government end all taxation of scholarships, fellowships, and student stipends for participation in research.

Finally, research-intensive universities can have a major effect on the development of scientific talent by educating inspiring teachers of precollege science and mathematics. PCAST, therefore, recommends that:

- research-intensive universities give greater emphasis to the education (including continuing education) of precollege teachers of science and mathematics; and

- the federal government provide scholarships or service-repayable loans to encourage talented students to attend research-intensive universities for careers as precollege teachers of science and mathematics.

Beyond the Horizon

This report is intended to address pressing problems that threaten the productive relationship between the federal government and research-intensive universities. We believe that the fundamental premises of this relationship are sound but that improvements are required for it to avoid deterioration and achieve its fullest potential.

We recognize the present time as one of tumultuous and profound changes in American society and in the world. The ending of the Cold War, the emergence of the European Community and nations of the Western Pacific Rim as economic powers, the changing demography of the American population, the ever increasing power of science and technology, and the growing awareness of new societal problems to which science and technology can be applied all require fresh and creative thinking of overall federal science and technology policy of which federal government-university relations are a part. While the current framework has served us well for four decades, it is far from obvious, as we move into the swifter current of the twenty-first century, that it will retain the validity it had when it was established in the middle of the twentieth century.

In our view, we must look beyond the immediate issues addressed in this report and conduct a broad national reexamination of the place of research and development in our national life — including its fundamental rationale, goals, organization, funding, and administrative mechanisms. The importance of generating new knowledge and new technologies and of educating scientists, engineers, and the general public for the twenty-first century demands no less. It is the intention of PCAST, drawing fully on federal and state government as well as private sector expertise and experience, to undertake such a reexamination in a subsequent report.

Renewing the Promise: Research-Intensive Universities and the Nation

Progress in the war against disease depends upon a flow of new scientific knowledge. New products, new industries, and more jobs require continuous additions to knowledge of the laws of nature, and the application of that knowledge to practical purposes. Similarly, our defense against aggression demands new knowledge so that we can develop new and improved weapons. This essential, new knowledge can be obtained only through basic scientific research.

Vannevar Bush
Science—The Endless Frontier

The Promise

With these words Vannevar Bush began his momentous 1945 report to President Truman on the role of science and technology in the postwar era. His message was straightforward: without scientific progress and advanced education, the nation could not hope for rapid increases in its standard of living, its economic well-being, the health of its citizens, and its national security.

Science—The Endless Frontier effectively promised the nation that science could yield enormous benefits if three conditions were met:

- That the United States, as a nation, make a substantial commitment to conducting basic research;

- That America's colleges and universities, both public and private, take on the responsibility of providing the nation with the requisite new knowledge and scientific and engineering talent;

- And that the federal government provide the funds to enable academic institutions to meet those new responsibilities.

Out of those conditions emerged a unique partnership between the federal government and America's colleges and universities. The public would invest in institutions of advanced learning and scholarship, and those institutions would in turn produce the new knowledge, understanding, and human capital that would fuel economic growth, improve public health, and strengthen national security.

What is a Research-Intensive University ?

No other nation has the advantage of such a large and diverse collection of world-class universities, each of which has developed strong capabilities in many fields of basic research, advanced education and public service. How many can be called "research-intensive" depends on the criteria applied — such as number of graduate students, national and international reputation, size of faculty, size and significance of their research and education programs, or amount of Federal research funding.

For purposes of this report, the precise number is not important. However, PCAST was curious as to what overall number of universities performed most of the research and educated most of the doctoral students. Using two criteria — namely, those that together produced 90% of the Nation's doctorates in science and engineering between 1981 and 1990 (the latest available year), and expended 90% of academic research and development funds from all sources during that decade — resulted in a total of 170 universities.

These universities are located in virtually every state. Some states with large populations, such as California, New York, Texas and Massachusetts, have several; others such as Michigan, Florida, Illinois and Wisconsin, have fewer but generally larger ones. About two-thirds are state institutions, and the rest privately-governed. Significant increases in state and federal funding for education and research during the 1960s, and again in the 1980s, led to the creation or expansion of many state universities — particularly as Southern and Western states grew in population.

The character of each research-intensive university depends on its individual history. They range, for example, from the large land-grant universities of the Midwest — with their teaching hospitals, agricultural extension services and sizable engineering schools — to mid-size institutions founded by religious organizations long before the Revolution, to smaller polytechnical universities that concentrate in a few fields. And beyond any total that could be generated using broad criteria, there are many other universities with one or several nationally-competitive research departments, developed in response to local or regional industry needs or some other feature unique to the institution's location.

The Promise Fulfilled

Vannevar Bush's vision ultimately led to the establishment of the largest and strongest scientific and higher education enterprise that the world has ever seen. This enterprise has fulfilled the promise of science to a greater extent than could possibly have been anticipated right after World War II. To cite a few of the most dramatic examples:

- Agricultural research conducted at our land-grant and other universities has served as one of the earliest models of a federal government-university partnership. The continuation of that partnership since World War II has led to productivity levels so great that two percent of the population of the United States produces enough food for our entire nation and large surpluses for export as well;

- Research in medicine and the biological sciences carried out in our universities and medical schools has led to vast improvements in the health of people around the world and has generated a new biotechnology industry in the United States;

Figure 1
Distribution of National R&D Funding by Source, Performer, and Character of Work, 1991

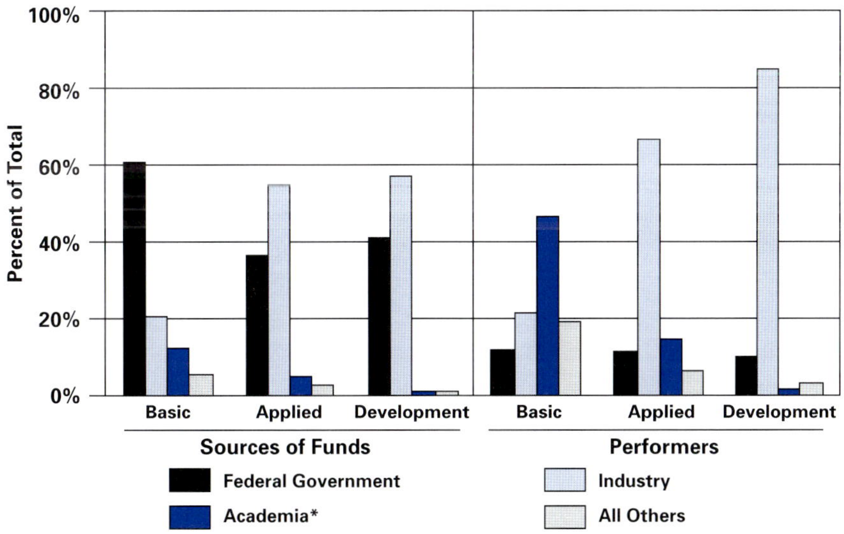

* For performers, includes university Federally Funded Research and Development Centers (FFRDCs)

Source: National Science Foundation

Diversity is a prominent characteristic of the U.S. R&D system. In no other nation do so many different sponsors foster discovery and innovation in so many different disciplines within so many different institutions. Yet, amidst this diversity, basic research is significantly dependent upon a single class of relationships — those where a federal agency is the sponsor and a university is the performer. During 1991, of the $23.5 billion expended within the United States for basic research (16 percent of all R&D expenditures that year), the federal government provided over 60 percent of the funds, while universities carried out almost half (47 percent) of the work.

- University research on the structure of matter and the properties of materials provided the key foundations for a multibillion-dollar microelectronics industry, which in turn has fueled computer and telecommunication industries and has changed the nature of society throughout the world;

- Basic materials research has also led to critical advances in aerospace and other technologies. These, together with electronic and computer technologies, have been among the resources with which the United States buttressed its national security during more than four decades of the Cold War;

- Social scientists at our universities have provided us with the fundamental concepts that underlie the survey research and economic analyses on which business institutions rely for their marketing and financial planning and on which federal, state, and local governments depend in making public policy;

- Many technologies flowing from fundamental university research have made possible a whole new era of entertainment, continuing education, and public information access;

- Fundamental advances in physics, chemistry, and biotechnology have spawned a revolution in environmental science and technology. For example, rather than simply dispose of hazardous wastes, the focus is now on reduction of waste at the source and on reclaiming valuable components of industrial and other waste streams to yield environmentally benign residues; and

- University research has provided us with much of our dramatically improved understanding of such natural phenomena as earthquakes, hurricanes, and other severe weather patterns. That knowledge has served to reduce the loss of lives, improve buildings, bridges, and other structures, and mitigate major economic losses — in short, to prevent natural hazards from becoming national disasters.

Basic scientific research and the associated development of scientific talent are only two agents of progress. Many kinds of institutions — public and private, industrial, and financial as well as scientific and educational — also contribute to our economic growth and well-being. But, over the past four decades, the American public has placed both its faith and its tax dollars in our colleges and universities so that the new knowledge, technological advances, and human talent that these institutions are uniquely capable of producing would in turn generate wealth, health, and security for the nation.

The record of those four decades is clear: the promise was more than fulfilled, the faith of the public was rewarded, and the investment continues to pay off.

THE GROWTH OF THE ENTERPRISE

The federal research enterprise in America has become large and very complex, involving more than twenty federal agencies, each with its own mission, managing hundreds of individual programs. More than one hundred and fifty universities are at the center of that enterprise. Because most faculty at those institutions devote considerable efforts and resources to research as well as teaching, the universities are referred to in this report as "research-intensive universities." They are dispersed over virtually every state in the Union and exhibit considerable diversity in size, character, and governance. The interrelationships between the federal government and those universities loom large for both partners, although this is but one component of a complex array of university activities and federal research programs.

Federal support of university- and college-based programs has grown substantially since World War II. This support has focused on student aid and the sponsorship of research activities, particularly basic research. Overall, the U.S. system of research-intensive universities is roughly three times the size it was thirty years ago: in enrollment and degree production at all levels, in numbers of faculty, and in numbers of research staff. The capacity of our universities to produce doctorates is well over ten times what it was in the immediate postwar years.

Figure 2
Federal Expenditures for Research in Total and for University-Based Research

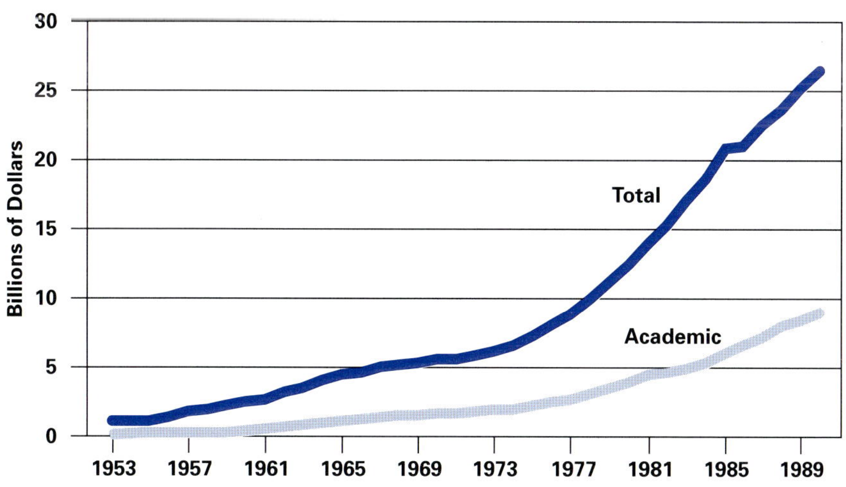

Source: National Science Foundation

The federal government has been and continues to be the principal source of funding for the substantial expansion of U.S. research activities that has been underway since the end of World War II and especially since the mid-1970s. Federal funding for university-based research has constituted a significant fraction of this investment throughout the last four decades, reaching approximately $9 billion or almost 35 percent of the total in 1990.

In real (inflation-adjusted) terms, federal support for university-based research is now about five times what it was thirty years ago. Although federal support is still concentrated in a relatively small number of institutions, over three decades more institutions have become larger partners in federally supported research and development (R&D). In the postwar period scarcely a dozen institutions received half the federal funds granted to universities. At present, over thirty universities account for that fifty percent share.

In sum, the implementation of Vannevar Bush's 1945 vision has led to the unprecedented growth of research-intensive universities throughout the nation.

Signs of Stress

The strength of the enterprise and the great value of its contributions to the nation notwithstanding, the relationship between the American public and institutions of higher learning is showing serious signs of stress. In the current public debate on research and education in American universities and colleges, the following points are frequently heard:

- Students and their families are concerned that the quality and value of the education provided by colleges and universities are not commensurate with its rising cost;

- Faculty and administrators express deep concern that, despite a steady and substantial growth of government support, resources for basic research and advanced training are falling far short of what is required either to permit exploitation of the myriad opportunities that have been opened up through research successes in recent decades or to enable our growing national pool of highly trained investigators to pursue their scientific careers;

- Both the university world and the tax-paying public have been disturbed by isolated but sobering incidents of faculty and institutional misconduct. This has diminished the nation's confidence in academic institutions;

- Various members of the higher education community have described the relationship of their institution to the federal government, particularly as it pertains to the support of university-based research, as increasingly adversarial; and

- Faculty and administrators are concerned that the federal government is demanding more applied research as a visible contribution to the nation's economic competitiveness.

VARIOUS MEMBERS OF THE HIGHER EDUCATION COMMUNITY HAVE DESCRIBED THE RELATIONSHIP OF THEIR INSTITUTION TO THE FEDERAL GOVERNMENT . . . AS INCREASINGLY ADVERSARIAL.

These points serve to underscore some of the diversity and multiplicity of partners in the nation's large and complex research and higher education enterprise. Those partners — students and parents, federal and state governments, foundations and industry, faculty and scientific communities — have ever increasing expectations about what the research-intensive universities can provide: more research, more education, more service to local industry and communities. At the same time, all those who have a stake in the enterprise — both inside and outside the universities — have become unsure that their expectations can be met. In a sense, the enterprise may have become the victim of its own success. As Hanna Gray, president of the University of Chicago, suggested in her 1992 keynote address to the American Association for the Advancement of Science,

> *Our universities have arrived at a stage of maturity burdened by too many tasks . . . and too great a confusion of expectations, by the consequences and distortions of excessive growth, . . . and by the illusion that comprehensiveness is necessary for institutional distinction.*

Ever Increasing Expectations

World War II left the United States in a position of economic preeminence with respect to both its defeated adversaries and its devastated allies. It also left us in a newly adversarial position with respect to the Soviet Union, the central and eastern European nations that the USSR had come to dominate, and its then ally, China. Much of the rest of the world was emerging from a long colonial era. The national security imperatives of the Cold War — particularly the aerospace and electronic requirements of defense — were one of the engines that drove much of the federal government's commitments to R&D for four decades. Other important engines have been commitments in the area of health and the needs for an assured energy supply and for a more highly trained labor force.

With the remarkably swift end of the Cold War and major political shifts in the world, the long-term national security needs of the United States cannot be easily defined today. What is likely is that uncertainty will become a feature of a more fluid international system. Preparedness in the face of uncertainty is even more complex and ultimately more technologically demanding. If anything, technological advances will be more important than ever before in ensuring our national security needs.

Figure 3

Increasing Dispersion of the Nation's University-Based R&D Activity as Shown by Expenditure Distribution in Four Selected Years

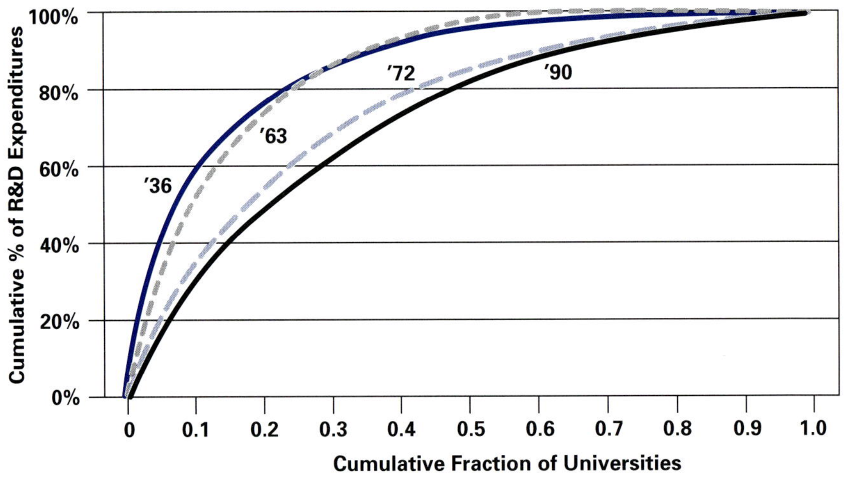

1936: National Resources Commission; 1963: Congressional Hearings: 1972, 1990: NSF Surveys

Source: National Science Foundation

The number of universities participating in the national R&D effort has been growing over the last several decades. As a consequence, the most research-intensive universities have accounted for a progressively smaller share of expenditures for university-based R&D. The changes in recent years, while consistent with the long-term trend, nevertheless are modest compared to those that occurred between the early 1960s and the early 1970s.

Although the Cold War, with its complex set of implications for the nation's R&D programs, is no longer center stage, new developments, particularly in the international economy, are presenting new challenges for the nation. The industrial economies of Japan and Germany as well as those of major Western European countries have been rebuilt, largely on the basis of technology-intensive strategies. Former colonial and quasi-colonial economies on the Pacific Rim — South Korea, Taiwan, Hong Kong, and Singapore — have surged ahead with export-led strategies rooted principally in the electronics industry. All of these countries, as well as others — such as China — on the horizon, will continue to pose a powerful economic and scientific challenge to the United States.

PCAST IS CONCERNED THAT AMERICA'S UNIVERSITIES ARE NOT AS WELL POISED AS THEY SHOULD BE TO ASSUME THE RESPONSIBILITIES OF THE FUTURE.

Paralleling the widespread economic and population growth of the past three decades has been the emergence of global environmental concerns. By their very nature these can only be addressed on a multinational basis — both in terms of a scientific understanding of complex physical, biological, and human phenomena and in terms of policy responses to environmental threats. This is another of the new R&D challenges before us. Finally, we are approaching a new frontier in the area of the life sciences that provides many new opportunities for both the relief of human suffering, the further advancement of world agriculture, and the development of important new industries.

As America looks to the future, it is clear that the nation's dependence on its colleges and universities for creating new knowledge and training new talent will be greater than at any time in our history. Over more than four decades our research-intensive universities have shouldered a weighty responsibility for the nation. Reflecting a unique American belief that higher education and research are inextricably linked and synergistic, we have made equally unique demands on our research-intensive universities — both for new knowledge and for young minds trained to use it in a creative and innovative fashion. No other nation has made comparable demands on their universities.

PCAST is concerned that America's universities are not as well poised as they should be to assume the responsibilities of the future. PCAST is also concerned that the American public no longer has the confidence in universities that they will need if they are to meet those responsibilities. This unhealthy situation puts the nation's future in jeopardy.

PCAST has sought to identify the weak points that threaten the otherwise strong web that has bound together for many decades the federal government and the tax-paying public with the research-intensive universities. In the six sections that follow, PCAST presents its findings and recommendations for restoring the relationship.

I. A Matter of Limits: Adapting to a New Resource Environment

"We are in the paradoxical situation of being unable to support adequately all of the valuable scientific work we are now capable of doing, while at the same time we are generating the capacity to do more."

Robert M. Rosenzweig, President, Association of American Universities
July 1992 statement to PCAST

"America's research universities today rest on unstable and shifting ground."

Charles Vest, President, Massachusetts Institute of Technology
June 1992 statement to PCAST

ADVANCING THE FRONTIERS OF KNOWLEDGE IS NOT, AS IT ONCE MAY HAVE BEEN, A MATTER OF INTELLECTUAL LUXURY. IN AN ERA OF RELENTLESS GLOBAL ECONOMIC COMPETITION, IT IS A NATIONAL IMPERATIVE.

Advancing the frontiers of knowledge is not, as it once may have been, a matter of intellectual luxury. In an era of relentless global economic competition, it is a national imperative. The United States must continue, as a nation, to invest in fundamental research. Because the ultimate applications that might derive from fundamental research are largely unforeseen, there are no easy answers to such questions as "how much basic research is optimal?" or "in what specific areas of science should investments be concentrated?" Expenditures in basic research are an investment with returns that, although unknowable in advance, have proved enormous in the past and are likely to be even greater in the future.

A realistic assessment of the challenges facing America in the next decades indicates that the key issues before us will be different from those of the past. In particular, unless we achieve more robust and sustained levels of economic growth we will face some very difficult choices. In such an environment, all public resource commitments will receive new levels of scrutiny, and the university-based system of research and education is not likely to expand for the indefinite future. It is probably unreasonable to expect that the system will continue to grow as it did for periods in the 1960s and 1980s. Given the resources that the nation is willing and/or able to devote to this enterprise, the system may already have exceeded its steady state capacity.

The financial stresses already experienced by both our institutions of higher learning and their many different supporters — students and their families, the federal government, and state and local governments — are only in part a reflection of ongoing economic cycles. These short-term conditions should not obscure longer-term challenges brought about by relatively slow growth in family income, intense international competition, and rising costs of cutting-edge education and research.

Quite apart from particular economic conditions, the nation has to contend with two sets of cost issues. The first is that the system contains two built-in growth factors. Almost all research breakthroughs open multiple opportunities that can be pursued, each holding the promise of new technologies, new industries, new conquests over disease, and new tools for improving the human condition. These developments raise national productivity, wage rates, and the cost of attracting talented investigators to the research enterprise. Furthermore, the system has a kind of "natural population growth." Each university scientist usually trains many new scientists, several of whom now typically remain in the university system.

The second cost issue pertains to inflation — not only that which affects the economy as a whole, but "sophistication inflation." Each scientific advance requires more complex and sophisticated techniques, instrumentation, and facilities than did the last. Each successive step forward is a longer step but a more costly one as well.

The cross-pressure of expanding opportunities for investment in research and constrained resources poses a risk for the United States: spreading our resources too thinly across an array of highly trained investigators and research-intensive universities. The relentless quest for an ever broader range of activities will inevitably destroy the most important aspect of American higher education and research — quality. We cannot afford to allow higher education to decline to the levels of mediocrity that now characterize much of our precollege education.

IT IS PROBABLY UNREASONABLE TO EXPECT THAT THE SYSTEM WILL CONTINUE TO GROW AS IT DID . . . IN THE 1960S AND 1980S.

Most of our research-intensive universities aspire to excel in all or most areas of scholarship and education. As worthy as those aspirations might seem, they are likely to be ill-advised. They cannot be fully realized in an era of relatively constrained resources. Our research-intensive universities must adopt more highly selective strategies.

Even for those who still believe in optimistic scenarios, an important observation applies. Those institutions that have maintained the highest standards of quality by being selective in their investments in faculty and physical plant are the ones that will be in the best position to compete successfully for and make effective use of any significant new resources that might become available.

Recommendation One: Adapt Quickly and Responsibly to a Constrained Resource Environment

So that research-intensive universities may continue to function productively in an environment of limited resources, PCAST urges them to:

> adopt more highly selective strategies based on a realistic appraisal of the future availability of resources and a commitment to meet world class standards in all programs that are undertaken.

Such strategies will require universities to:

- eliminate or downsize some departments and specialties within departments rather than sustain less than world-class activities in every area of science and engineering;

- collaborate with nearby institutions (academic, industrial, and governmental) in sharing instructional and research facilities and programs, with the aim of conserving limited resources;

- build facilities or programs only where there are strong long-term prospects of sustaining them. The expectation that, somehow, new resources will become available to sustain initiatives must be viewed with skepticism. Without long-term sustainability, initial investments result in squandered resources; and

- develop permanent institutional mechanisms for strategic planning that will foster a balance between activities and resources and among teaching, research, and other missions that are commensurate with society's values and the needs of the universities.

SPREADING OUR RESOURCES TOO THINLY . . . WILL INEVITABLY DESTROY THE MOST IMPORTANT ASPECT OF AMERICAN HIGHER EDUCATION AND RESEARCH — QUALITY.

PCAST also urges federal agencies and other supporters of research-intensive universities to:

- refrain from encouraging universities to embark on new research, new education programs, or the building of facilities when there is little or no long-term prospect of sustaining them; and

- refrain from developing or implementing research or education programs that would increase the net capacity of the nation's research-intensive universities.

II. A Matter of Education: The Importance of Teaching at Our Universities

> *Conducting research has become such an overwhelming focus on today's campuses that those professors who still manage to teach more than a few hours a week are actually looked down upon by their peers, to say nothing of the negative effect teaching has on chances for tenure, pay and promotion.*
>
> Representative Patricia Schroeder
> Chairman, Select Committee on Children, Youth and Families

> *The best teacher in the world is known only to the perimeter of his campus, while a mediocre researcher is known around the world.*
>
> Norman Hackerman, former president of
> the University of Texas at Austin and of Rice University and
> former Chairman of the National Science Board

To meet a wide range of its future needs, the Nation will have to educate a higher percentage of its citizens in science and technology. Those needs include:

- filling demanding positions in competitive, high-technology manufacturing industries such as computers, biotechnology, and medical instrumentation;

- finding solutions to a wide range of problems in such areas as environmental control and disease prevention;

- making sound decisions in technically-based spheres of corporate management, legal affairs, and public policy; and

- participating effectively as citizens of a democratic society, by having the ability to understand the principles of technologically-oriented public policy issues.

But at the very time that highly-trained scientists, engineers, and technicians and scientifically literate lay persons are more important than ever to our society, declining numbers of high school students are interested in science and technology, and most have inadequate precollege preparation in these areas.

In addition, the traditional model for producing scientists and engineers — where one completed a formal college or graduate school education and then moved on to career activities having completed education — if it was ever true, is no longer valid. If the system does not adapt to massive changes in the demographic composition of the student body (which is older on the average, culturally and ethnically more diverse, and includes many individuals attending college part-time) and does not take advantage of relatively new post-secondary institutions such as community colleges, opportunities may be foreclosed to some talented

The U.S. Higher Education System

Higher education is a major, complex and diverse enterprise in the United States. Overall, it comprises more than 3,000 colleges, universities, specialized institutions and professional schools, employing a total of 750,000 faculty and serving about 12.4 million students.

Several types of institutions play varying roles in educating students for mathematics, science, medicine and engineering. Many of the 1,250 two-year colleges have important missions in technician training, especially to meet the needs of local industry — and a sizable and expanding role in preparing students for transfer to baccalaureate institutions. There are about 1,270 institutions with programs primarily at the bachelors'- and masters'-degree levels. They annually award about 180,000, or 55%, of the baccalaurates, and about 40% of the 60,000 masters' degrees, in science and engineering.

Another way of viewing higher education institutions is by the student populations they serve. More than a hundred historically black colleges and universities, some of considerable distinction in research, education and medicine, have evolved from segregationist beginnings in Southern states. Also, as the demography of the U.S. population changes, a growing number of colleges and universities in several areas of the country now have student bodies comprising sizable numbers of Hispanic-Americans and Asian-Americans.

Faculty at those institutions and at hundreds of others contribute to the Nation's research effort; about 1,100 colleges and universities annually report at least some spending on research. A small group of highly selective liberal arts colleges send a relatively larger proportion of their students on to the doctorate than do most four-year institutions; indeed, faculty at these "research colleges" regularly win nationally-competitive funding for their own research.

In fields other than science and engineering, almost all four-year degree holders have completed two or more science courses and two or more mathematics courses. Many of them will become pre-college science or mathematics teachers, and many others will become leaders in business or government who make decisions involving scientific or technological matters. And most are (or become) citizens who will exercise many choices during their lifetimes about public issues having substantial scientific or technical content.

persons not on the traditional path. Education must be viewed as a continuing, life-long component of the life of each of our citizens, and an attractive, accessible means of achieving life-long education must be provided throughout the nation.

Improved teaching and learning of science and technology are needed throughout the system. All institutional levels — elementary schools, junior high schools and high schools, two-year colleges, undergraduate colleges and universities, and doctoral institutions and those involved in continuing education — have a role and a stake in providing a high-quality education in science and technology to all of their students. Research-intensive universities have a particularly powerful role as a consequence of their extensive research programs and associated activities with industry and government and their responsibilities in preparing the next generation of teachers.

Figure 4
Number of Academic Institutions by Type

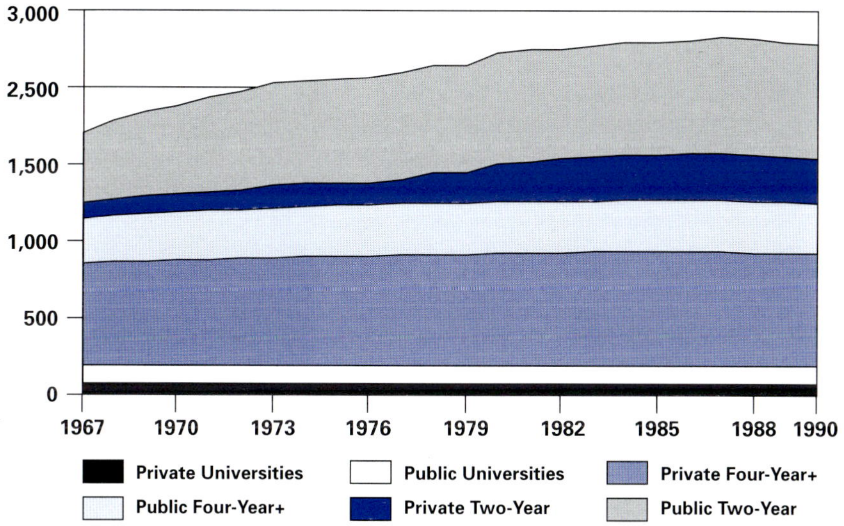

Source: National Science Foundation and Department of Education

The U.S. system of higher education could not possibly have kept pace with the enrollment increases without significant structural change; and that change was well underway by the mid-1960s. The most prominent feature was the creation of about 1,000 new two-year colleges. The continuing challenge with respect to science and technology is to help ensure that these colleges provide the educational opportunities — both directly and through affiliations with four-year colleges and universities — that will prepare students for science and technology careers.

"Universities" refers to institutions typed by the Carnegie Foundation for the Advancement of Teaching as "research" and "doctoral". "Four-year+" refers to Carnegie's "comprehensive" and "liberal arts" institutions.

Figure 5

Higher Education Enrollment* by Type of Institution

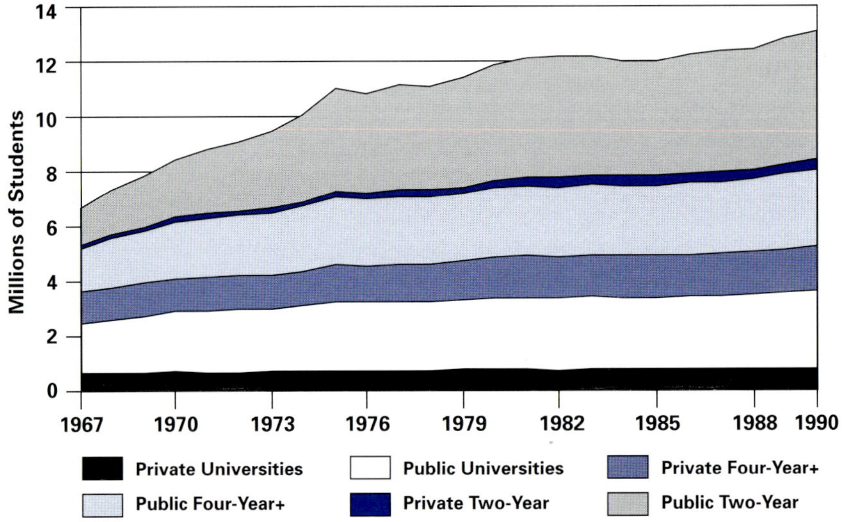

* Includes part-time students. Excludes highly specialized schools providing religious, psychiatric, chiropractic, etc. training.

Sources: National Science Foundation and Department of Education

The decline in college and university enrollments that inevitably accompanied World War II quickly reversed during the post-War years as former members of the military, many aided by the G.I. Bill, entered or reentered higher education. However, by the mid-1950s, when most of the cohort of returning veterans had passed through the system, the level of enrollment once again became primarily a function of the number of traditional "college age" persons in the population — i.e., the 18-21 year-olds.

This pattern held until the mid-1960s, after which time enrollment in undergraduate programs began growing much more rapidly than traditional "college age" population and then continued growing, albeit more slowly, despite the decline in that age group as the "baby boom" cohorts successfully achieved adulthood. Since the mid-1960s, the number of students enrolled in U.S. institutions of higher education has grown by a factor of four — i.e., from about 3.5 million in 1960 to about 13 million in 1990. This was the result of increasing participation of many different age groups, especially students over 25 years of age, as well as women and minorities in general.

However, many higher education institutions, including research-intensive universities, have come under fire for seeming to turn away from education, especially of undergraduates, toward research and other activities. Parents and students complain of:

- too little direct contact between undergraduate students and senior faculty, both in and outside of the classroom;

- too much reliance on persons other than faculty members for educational counseling;

- too many early courses taught by graduate teaching assistants — some of whom are ill-prepared to teach, some not fluent in English, and some exhibiting hostile attitudes toward women and minority students;

- too many course offerings, including some that appear duplicative or that reflect very narrow faculty interests peripheral to students' needs for an intellectually integrated education;

- years of continuing tuition increases, severely pressuring a family's capability to afford higher education — particularly given recent economic conditions; and

- failure to orient students to the full range of intellectual and career possibilities of the world into which they will graduate.

America's educational needs are greater than ever. It is critical to the nation's future that universities reemphasize their educational mission and apply their unique resources to society's educational requirements. PCAST believes that increased attention to educational activities need not drive up the costs of education and need not be at the expense of the net research output of truly valuable new knowledge from the research-intensive university system as a whole.

Reemphasizing education will require improving teaching performance and raising teaching productivity, as difficult as those may be to evaluate and measure. While experience has demonstrated clearly that prior research performance is the best available indicator of future professional productivity, it is fundamentally more difficult to measure teaching performance. For example, evaluations of teachers by students five years after graduation are typically dramatically different from evaluations made at the time courses are taken.

Recommendation Two: Reemphasize Teaching

Each research-intensive university should review, in a searching and comprehensive manner, the nature and quality of its teaching programs — particularly in science, mathematics, and engineering — with a view to improving teaching performance and productivity. PCAST in no way implies that performance and productivity problems are limited to those fields; rather our experience and expertise limit our consideration to them. More specifically, each research-intensive university should develop policies and programs to:

- increase direct senior faculty involvement in teaching at both the undergraduate and graduate levels and in counseling of students;

- balance the contributions of teaching and interaction with students with those of research and public service in evaluating and rewarding faculty;

- place less reliance on graduate teaching assistants and ensure that they understand their subjects, are better prepared with regard to teaching methods, are able to express themselves well in English, and, like all faculty, are able to provide a supportive environment for women and minority students;

- increase efforts to involve undergraduates in hands-on frontier research; and

- place greater stress on educating scientists and engineers in key foreign languages.

Even as universities reemphasize teaching, they and their patrons working together must keep tuition and educational costs from rising faster than the income of the average American family.

In addition, the research-intensive universities should:

- develop new pedagogies, including technology-based teaching methods, for undergraduate teaching. These new pedagogies should be aimed as much or more at non-science majors as at those intending to pursue careers in science and technology;

- help two-year institutions improve curriculum quality and develop agreements about transferability of credits;

- where appropriate resources and talents exist, assist with national, state, and local efforts to redesign and revitalize precollege education, especially in science and mathematics; and

- give special attention to undergraduates who intend to become precollege teachers of science and mathematics. These individuals must have a thorough grounding in those topics.

The federal agencies should ensure that their programs encourage universities to reemphasize education rather than discourage them — even inadvertently — from taking the measures recommended above. In particular, federal agencies should:

- provide incentives for outstanding undergraduate and graduate teachers in the fields of science and technology, such as awards and special national recognition for their accomplishments; and

- develop or expand sustainable programs that would assist research-intensive universities in meeting the objectives described above.

III. A Matter of Public Trust: Restoring Confidence in Our Universities

Cutting through all of the "excellence" and "quality" rhetoric reveals one very clear point: the focus in higher education today is on research, not teaching. This fact has not been lost on the professors. If you don't believe me, go ask one yourself. However, don't look for a professor in a classroom; it's unlikely you'll find one.

Representative Patricia Schroeder
Chairman, Select Committee on Children, Youth and Families

"Universities have been over-responsive to those who seem to think that they should carry out every function and address every concern that might be of interest to citizens in general. They need to return to the criterion that measures what they can do, and do well, that other institutions cannot do, or do as well, and stick to their own special purpose, or it will be lost."

Hanna Gray, President of the University of Chicago
Keynote address to the American Association for the Advancement of Science, 1992

PUBLIC CONFIDENCE IN UNIVERSITIES IS ERODING.

Public confidence in universities is eroding. Although studies show that the economic value of an advanced education has increased substantially in the last decade, there is nevertheless a growing concern that tuition and related costs are rising too quickly and that the teaching programs of the research-intensive universities should receive more attention. Students, parents, and legislators are raising concerns that those institutions have shifted their attention too far toward performance of research and away from dedication to students. In the current economic environment, continuing tuition increases leave middle-class parents wondering how they will be able to afford to send their children to an institution of their choice and whether they will receive full value for what they invest.

Revelations that a few universities, in recovering indirect costs of research, billed the government for items having little, if any, relationship to research efforts have further served to undermine public trust, as have a few widely publicized instances of scientific misconduct. More generally, some outside the academic community have lost faith in the vision and commitment of university faculty and administrators, in their management of universities, and in their use of the resources currently available to them. Exacerbating these problems have been responses from some of the individuals and institutions involved that were interpreted by the public as arrogant.

The notion of a university as a community of scholars having shared interests in certain critical tasks, such as undergraduate education, is under siege from inside universities themselves. Some members of the academic community — at times spurred by their reassessment of our history and culture — act as if they have lost faith in the continuing vitality of such core values of the scholarly enterprise as free inquiry, tolerance, scholarly integrity, verifiable ways of knowing, and open communication of new ideas. Current "politically correct" activities on many campuses attack the fundamental values and foundations of the modern university. Some faculty pursue their own work without regard for their institution as a whole or for the coherence of its teaching curriculum.

Factors that undermine public trust can be placed in three general categories. Some, such as fabrication of research results and misuse of funds, however isolated in their incidence, are universally acknowledged as grossly inappropriate and illegal. Some, such as banning of "politically incorrect" speakers or teachers, undermine some of the deepest values of academia. And some, such as debates about the balancing of efforts among research, teaching, and community outreach, may reflect important disagreements about priorities of missions and resources.

Figure 6

Undergraduate Enrollment as a Percent of Total Enrollment in Research-Intensive Universities

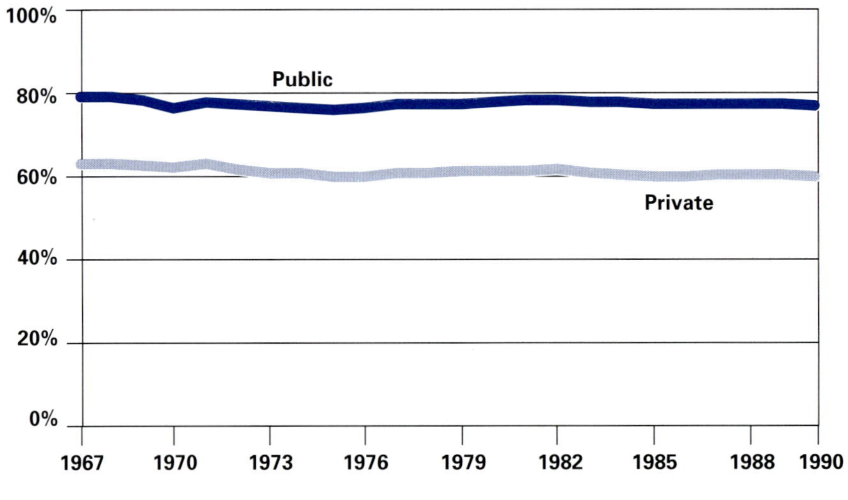

Source: National Science Foundation

Undergraduate education is a significant function for the research-intensive universities. Bachelors-degree candidates have accounted consistently for about 60 percent of the enrollment in private research-intensive universities and about 80 percent of the enrollment in public research-intensive universities.

Research-intensive universities are very complex institutions with many outputs and many patrons, each of whom tends to expect the output it funds to have the highest priority and to be the most heavily subsidized by others. Students (and their parents) expect a high-quality education, including significant personal contact with the senior faculty that give the institution its reputation. State governments expect state-supported universities to offer a full range of graduate and undergraduate courses, at low tuition, to a large proportion of the state's high school graduates. Local industry and government expect to be able to hire graduates that are ready to work in their companies and agencies. Universities with medical schools and teaching hospitals are expected to offer high-quality, affordable health care to the public. Agricultural and forestry interests depend heavily on university extension services. Companies, foundations, and government agencies that help fund research expect top-rank work, well-educated graduates, and, increasingly, efforts to ensure that research results are used to generate industrial advances and jobs.

RESEARCH-INTENSIVE UNIVERSITIES ARE VERY COMPLEX INSTITUTIONS WITH MANY OUTPUTS AND MANY PATRONS, EACH OF WHOM TENDS TO EXPECT THE OUTPUT IT FUNDS TO HAVE THE HIGHEST PRIORITY AND TO BE THE MOST HEAVILY SUBSIDIZED BY OTHERS.

Having many missions and sponsors tends to pull research-intensive universities in several conflicting directions at once. Various sponsors demand that they:

- be both more business-like (e.g., focus on the "bottom line") *and* less business-like (focus on maintaining ideals and not "merely" on the bottom line);

- do both more scientific research (e.g., to help with economic competitiveness) *and* less research (e.g., more teaching and public service);

- both increase quality and quantity of services *and* cut costs; and

- be both the guardians *and* the critics of our cultural inheritance.

Each institution has to confront its unique set of conflicting demands originating with its unique set of constituencies. Whatever its approach, however, each institution must focus on the fundamental issue of public trust and confidence. Without those and the ensuing public support, our nation's research-intensive universities will not be able to meet the nation's needs.

Recommendation Three: Restore Public Trust and Confidence

In our view:

> the university community and its patrons, including federal agencies, must act in ways designed to preserve the core values that underlie the scientific and educational enterprise — free and creative pursuit of ideas and their implications, scholarly rigor, trust in rationality and in verifiable forms of knowledge, and belief in the benefits of the synergism of research and teaching as investments in the nation's future. Intolerance for alternative ideas, or absolute insistence on "politically correct" behavior, represents a kind of dry rot that will, if allowed to remain, undermine the nation's institutions of higher education.

In addition to reemphasizing education (see Section II), we urge colleges and universities to:

- establish effective procedures to eliminate fabrication, falsification, and plagiarism in scholarly work, and to eliminate fraud and waste in the administration of that work.[1]

PCAST cautions, however, that excessive efforts to anticipate and eliminate all potential problems can lead to bureaucratic strictures that undermine or stifle scientific creativity.

---

[1] A valuable discussion of this subject can be found in the recent report of the Panel on Scientific Responsibility and the Conduct of Research of the Committee on Science, Engineering, and Public Policy, entitled *Responsible Science: Ensuring the Integrity of the Research Process*, Vol.I, National Academy Press, Washington, DC, 1992.

IV. A Matter of Wise Investments: Federal Support of University-Based Research

While watching Michael Faraday at an experiment, British parliamentarian William Gladstone was said to ask 'Of what use is such a discovery ?' Faraday is said to have replied: 'Why, sir, you will soon be able to tax it !' A more recent British politician, Margaret Thatcher, recently noted that 'Although basic science can have colossal economic rewards, they are totally unpredictable. And therefore, the rewards cannot be judged by immediate results. Nevertheless, the value of Faraday's work today must be higher than the capitalization of all the shares on the stock exchange.'

recounted by Thomas Everhart,
President of the California Institute of Technology
July 1992 statement to PCAST

Short-term constraints on R&D resources as well as the likelihood that resources will be constrained well into the future have fueled a healthy national debate over priority setting in all sectors. The federal government — both in the Executive Branch agencies and in Congress — state and local governments, universities, industry, scientific associations, and foundations have all debated and searched for the optimal strategies for expending public and private resources on R&D.

The choices to be made are many: within fields of science; among fields of science; between "pure" research that has no obvious immediate application and "strategic" or "thematic" research that has some expectation of yielding short-term practical applications; among basic research, applied research, and development; between "big" science and "small" individual investigator science; between research itself and the facilities and instrumentation required to conduct research; between government laboratories and universities.

INTOLERANCE FOR ALTERNATIVE IDEAS, OR ABSOLUTE INSISTENCE ON "POLITICALLY CORRECT" BEHAVIOR, REPRESENTS A KIND OF DRY ROT THAT WILL, IF ALLOWED TO REMAIN, UNDERMINE THE NATION'S INSTITUTIONS OF HIGHER EDUCATION.

None of these choices is easy, and this report will not add one more set to the many contending options already on the table. Rather, we stress a few essential principles:

- Our nation must continue to invest enough in basic research to sustain world class accomplishments in all major areas of science and technology. In those areas where U.S. activity does not define the frontiers, it must be sufficiently close to those frontiers that we can exploit discoveries without delay, wherever and whenever they are made;

- A healthy federal government-university partnership — particularly in basic research — must be maintained as part of the core of a successful U.S. science and technology enterprise. It has served the nation well since World War II and we have every reason to believe that it will do so in the future;

- The federal government must, in its interactions with universities at all levels, recognize that it is investing in institutions that simultaneously generate new knowledge and new talent, and not just procuring research results. The knowledge and talent generated may not always directly or immediately benefit a particular government mission, but it constitutes important national capital that will serve many different purposes;

- The federal government must pay the full costs of the university-based research it supports. To expect cost-sharing (except where it strengthens the research-teaching linkages or in other special circumstances) defeats the investment objective and may shift costs to students and their families. The universities, on the other hand, must commit to using all federal funds that support facilities and equipment for the maintenance of those important elements of their research infrastructure;

OUR NATION MUST CONTINUE TO INVEST ENOUGH IN BASIC RESEARCH TO SUSTAIN WORLD CLASS ACCOMPLISHMENTS IN ALL MAJOR AREAS OF SCIENCE AND TECHNOLOGY.

- Federal funds for research should be allocated through competition on a merit basis. There is no better method for ensuring quality and maximizing the dividends of our investment than for proposed research to be subjected to competition based on scientific merit;

    — There are currently two specific areas in which the principle of merit-based competition is not followed. One is in the support of basic research in federal laboratories. Though this research may be reviewed for its technical merit, it is rarely subjected to direct competition with research proposals from other institutions, such as universities. When the primary mission of a laboratory has changed, is not clear cut, is self-generated, or overlaps the missions of other agencies, such competition, especially with the research-intensive universities, should be the norm;

    — The other area that circumvents merit-based competition involves Congressionally earmarked facilities and projects. Such facilities and projects, which are often intended only to satisfy particular interests, do not constitute a good investment for the nation. Instead, they waste federal funds, undermine morale, and destroy the integrity of the merit review process. The practice must cease and must not be initiated or encouraged by universities; and

- Finally, we note that priorities, by their very nature, are very sensitive to changes in the environment in which they are established. Criteria on which the priorities are based should, however, be much more stable and remain as suitable bases for new priorities appropriate to changed environments. Much more attention should therefore be focused on the development of criteria that this nation accepts as reflecting its long-term goals and aspirations and, as such, form an appropriate basis for the development of priorities at all levels and under all conditions.

Figure 7

University-Based R&D Support by Selected Federal Agencies (FY 1990)

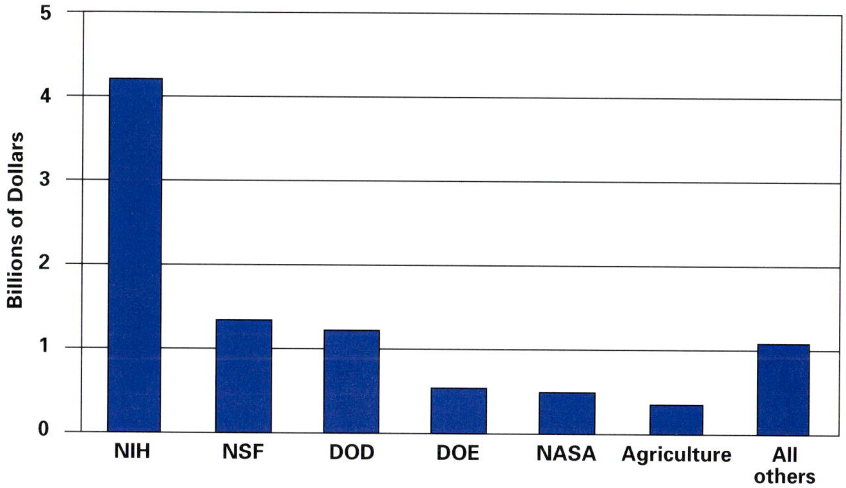

Source: National Science Foundation

More than 20 different agencies of the federal government fund academic R&D. Their individually unique program goals dictate their respective funding levels and associated spending priorities. The comparison of the top six agencies and the aggregation "all others" highlights both the importance of biomedical research in U.S. national priorities and the special emphasis that the National Institutes of Health places in investigator-oriented research and basic research in particular.

Recommendation Four: Adjust Federal Support of Academic Research

The basic approach:

- Federal agencies should view their grants in support of university-based research — even when undertaken in support of a narrow mission — as an investment in the nation's future, and not just as procurement of goods and services. The federal science and technology agencies have a collective responsibility to make those investments. The view that it is "another agency's responsibility, not ours" is not acceptable;

- Grants in support of research should be seen by federal agencies, furthermore, not only as investments in new knowledge but as investments in the nation's scientific talent;

- Federal agencies should review the balance between their intramural research and the university-based extramural research that they support with a view to maximizing the amount of research conducted at universities, where human capital is generated in tandem with intellectual capital;

- The underlying principle that federal agencies should follow in awarding research funds to universities is, except in special circumstances, to cover the full costs — both direct and indirect — of the research programs they sponsor. In turn, the universities must commit to use these funds in the fulfillment of agreed-upon objectives; and

- Except in special circumstances, such as in the case of professional schools, federal agencies should gradually withdraw from the practice of paying a portion of the faculty salaries guaranteed by universities. Tenured faculty should have their academic year salaries paid by the universities, to avoid artificial expansion of teaching faculties dependent on federal sources for institutionally guaranteed salaries.

THERE IS NO BETTER METHOD FOR ENSURING QUALITY AND MAXIMIZING THE DIVIDENDS OF OUR INVESTMENT THAN FOR PROPOSED RESEARCH TO BE SUBJECTED TO COMPETITION BASED ON SCIENTIFIC MERIT.

- In making awards, federal agencies should:

  — avoid undue specificity in stating expectations regarding the outcomes of research projects;

  — accord principal investigators maximum flexibility under the law with respect to the choice of proposed research goals, and actual approaches, methods, and use of resources;

  — make longer-term (e.g., three to five year) awards, whenever possible;

  — award more block grants to give established research groups stable, flexible support; and

  — be more willing to accept risk in supporting unconventional ideas, especially if proposed by investigators with a sound record of accomplishments. This would especially include cases of experienced investigators moving into new research fields. In such cases, more emphasis should be placed on their achievements and promise than on the particular details of their first research proposals in a new area.

Indirect costs:

- The Executive Office of the President should strive to ensure that the indirect cost portion of research awards meets both the requirements of modern scientific inquiry and the responsibilities that attend stewardship of public funds. To that end,

  — indirect cost policies should be refined to ensure that actual reimbursements cover only legitimate overhead expenditures;

  — indirect cost rates should be negotiated at levels sufficient to provide full reimbursement for those overhead expenditures — especially facilities-related costs — that should be allocated to the research sponsored by the federal government; and

  — all federal agencies should be required to honor the negotiated rates in full when making research awards.

▪ Universities should be required to set aside and use the facilities and equipment portions of indirect cost payments for maintaining, refurbishing, and renovating the physical infrastructure and equipment required for research.

Facilities and other infrastructure:

▪ The federal government should establish a temporary, nationally competitive, merit-reviewed program of grants for repair and renovation of the physical infrastructure of university-based research, such as buildings, specialized equipment, and computer networks;

▪ The program should operate on a 50-50 matching basis with non-federal funding. The size of the program should be commensurate with the repair and renovation needs. Recent estimates place those needs at more than $4 billion;

▪ That program should operate no longer than a few years, only until universities can bring their facilities to an acceptable level of modernity. Regardless of where the responsibility might lie for the current obsolescence of many university research facilities, the nation cannot afford to have that situation continue or deteriorate further. Beyond the catch-up period, however, universities would be expected to keep their facilities current on the basis of federal indirect cost support and other resources; and

▪ The competitive, merit-reviewed program should make funds available only to institutions that pledge to forego funds earmarked for award without such review.

FEDERAL AGENCIES SHOULD VIEW THEIR GRANTS IN SUPPORT OF UNIVERSITY-BASED RESEARCH — EVEN WHEN UNDERTAKEN IN SUPPORT OF A NARROW MISSION — AS AN INVESTMENT IN THE NATION'S FUTURE, AND NOT JUST AS PROCUREMENT OF GOODS AND SERVICES.

Less red tape:

Federal agencies should strive to ensure that the administrative requirements associated with research awards to universities facilitate scientific inquiry rather than impede it. To that end:

▪ funding agencies should authorize their program staffs to make small, short-term grants at their own discretion for particularly promising proposals where quick response would be especially advantageous, using streamlined application procedures and without external review; and

▪ the Executive Office of the President should take the lead toward achieving federal government-wide uniformity and eliminating unproductive administrative requirements by reaffirming its support for the goals of the Federal Demonstration Project and promoting full participation by the relevant agencies.[2]

---

[2] The Federal Demonstration Project (FDP) is devoted to improving efficiency in the administration of research grants. The continuing objective of the project is to identify and eliminate unproductive requirements without compromising stewardship of public funds. Ten federal agencies and approximately fifty universities participate in the project. FDP is the only organized, long-term effort aimed at streamlining the basic administrative relationships linking the agencies sponsoring research with the institutions that perform it. Outcomes of the FDP demonstrations usually consist of recommended changes to federal government-wide policies provided by the Office of Management and Budget.

Figure 8
University-Based R&D Support by Source of Funds and Type of Institution, 1990

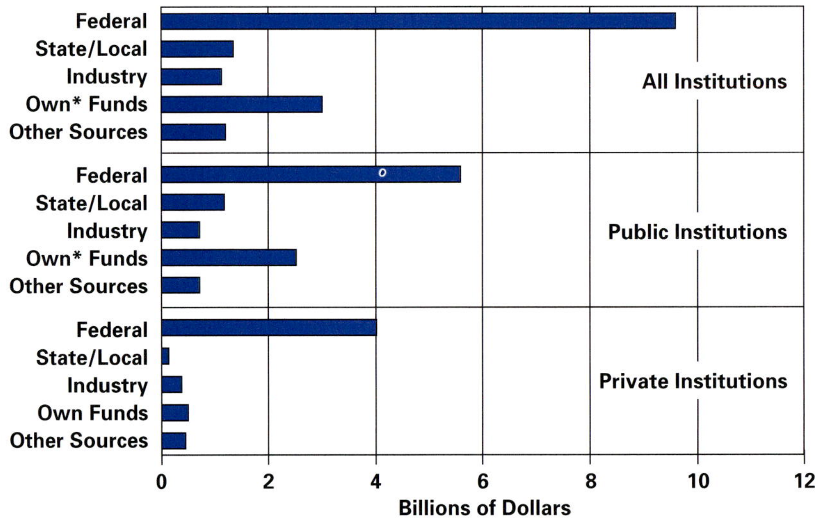

* Own funds for public institutions includes unknown amount of money from state sources.

Source: National Science Foundation

The federal government is the principal sponsor of university-based R&D. However, the federal share, which has essentially stable during the 1970s (near 67 percent), has been declining almost steadily since 1980, falling to 59 percent in 1990. The share financed by state and local government also shrank slowly throughout the 1980s while that financed from nongovernmental sources rose from 21 to 33 percent. Offsetting the contracting proportions provided by governments were the expanding proportions attributed to universities' own funds and industrial sponsors. The share of funding from all other sources (e.g., foundations) changed little from one year to the next, fluctuating around 7.5 percent.

Public and private universities exhibit some notable difference in the profile of their research support. The private universities are relatively more dependent on the federal government, whereas the public universities are more dependent on their "own" funds (see note underneath figure). The share of university-based R&D programs directly financed by state and local governments is higher for public universities than for private universities. In 1990, federal, state, and local governments together accounted for about two-thirds of university-based R&D at public universities and about three-quarters at private universities.

Federal Laboratories

Because federal support for research-intensive universities is affected by agency commitments to federal laboratories, PCAST believes there is now an urgent need to reexamine the roles of the more than seven hundred federal laboratories. As original missions have been accomplished, many of these laboratories have come to support basic research efforts lacking a clear relationship to mission objectives and in direct competition with research-intensive universities. These efforts typically have the benefit of superior resources, are not burdened by educational responsibilities, and are not subject to the same type of merit-review that ensures high standards of academic research. In some cases, new laboratories, in response to new missions, have been established when existing federal laboratories in other agencies are already setting the standards for activities in the fields covered.

Many of these laboratories continue to have appropriate and unique roles, including provision and operation of facilities beyond the scope of individual universities but invaluable to both university faculty and students. Under the impetus of the National Technology Initiative and similar programs, federal laboratories are forging new and productive linkages with U.S. industry. In the past, linkages between federal laboratories and universities were very strong, with a large flow of people back and forth among them and with resulting benefits in education and training. This flow has slowed, largely for bureaucratic reasons, and should be reestablished. It is appropriate to consider making all federal basic research support available for merit-based competition by universities, federal laboratories, or industry. Merit review in this case should include, as additional criteria, potential long-term contributions to economic well-being, national security, and education.

PCAST believes that a review of the federal laboratories similar to the present review of the research-intensive universities would be timely and would provide valuable input to the more general reexamination of the U.S. research and development enterprise that we discuss elsewhere in this report.

V. A Matter of Two Cultures: Universities and Industry

> *Senior management in industry must assure individuals who return to universities . . . that it is in the best interest of the corporation and the individual's career. This is not the case today. The university must be willing to take people who have not widely published . . . and are not recognized as the leading expert in a particular area. I believe both of these changes are possible. I believe the Nation will greatly benefit from this increased interaction.*
>
> William Spencer, President and CEO of SEMATECH
> July 1992 statement to PCAST

DESPITE RECENT GAINS IN BUILDING LINKAGES BETWEEN U.S. UNIVERSITIES AND INDUSTRY, THERE ARE STILL TOO MANY INDIVIDUALS IN BOTH SECTORS WHO HOLD TO NEGATIVE PERSPECTIVES, ATTITUDES, AND STEREOTYPES ABOUT THE OTHER SECTOR.

Two very different cultures have surrounded industrial research and university research. In industry, the drive for new products and processes sets the agenda in applied research and provides the context for fundamental research. University research is driven by a wide range of factors, some involving practical problems confronting society, but many curiosity-driven and associated with the pursuit of knowledge with no obvious relevance to immediate practical problems.

Some of the cultural differences have had the unfortunate effect of unnecessarily inhibiting full interaction between industry and universities. The notion that each sector had its own well-delineated and isolated role and that new knowledge would flow as rapidly as necessary and in one direction from the universities to industry is completely at odds with today's world.

Today the pressure of international competition has introduced a critical time dimension into the system. For the nation's economic interests, the issue is not simply how much new knowledge is being generated but also how fast it is being translated into economically and socially beneficial products and processes. This argues for a more deliberate effort to move information and, especially, people between universities and industry.

Over the past decade, substantial efforts have been made by federal, state, and local governments to foster greater and more effective ties between universities and industry. These efforts, which take the form of co-operative programs, research centers, and the like, should certainly continue. In addition, some scientific fields have developed in such a way that commercial applications derive more readily and rapidly from university-based fundamental research than was previously the norm. Biotechnology is the prime example, but a number of others could be cited.

The shrinking interval between fundamental research and industrial applications in certain areas also is serving to bring universities and industries together. Although these increasing linkages between the sectors have some potentially negative side-effects for universities, such as pulling faculty away from their teaching responsibilities or challenging strategic coherence at the institutional level, bringing universities and industry closer together in appropriate partnerships is, on balance, of enormous benefit to the nation.

Despite recent gains in building linkages between U.S. universities and industry, there are still too many individuals in both sectors who hold to negative perspectives, attitudes, and stereotypes about the other sector: new Ph.D.s who view taking a job in industry as "selling out" rather than following an academic calling; industry managers who are unwilling to send their best people to a university setting, even for a short time; faculty members who believe that their only educational mission is to train students for faculty positions and who channel their best students away from non-academic careers; industrialists who view university work as an intellectual luxury; academics who view industrial R&D as intellectually second rate.

The nation cannot afford to have this situation persist, and much more effort is required to overcome it. Even fundamental research that is not expected to yield short-term answers to industry's scientific problems can benefit from being informed by the technical concerns of industry. Conversely, U.S. industry should have the benefit of easy and immediate access to the new knowledge and new talent generated by universities. Exchange of personnel, at all levels, is the surest answer to these problems.

Figure 9
Bachelors Degrees by Broad Area of Study
(Thousands of Degrees)

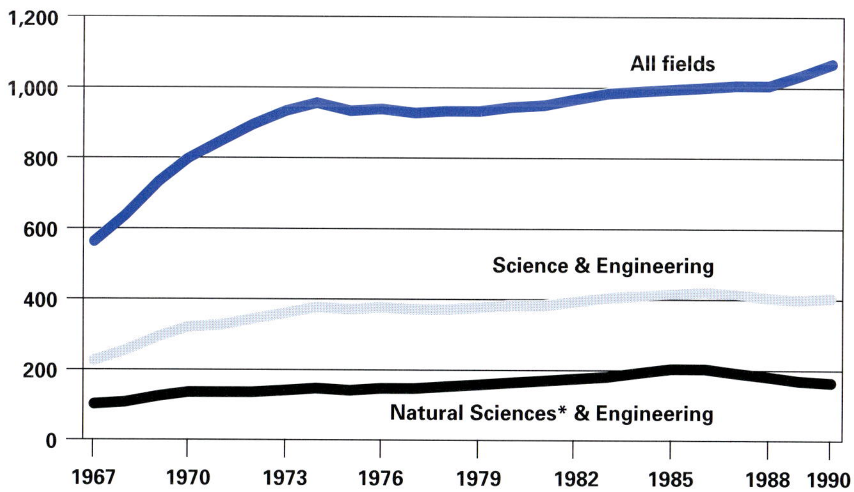

* Does not include medical and social sciences and psychology

Source: National Science Foundation and Department of Education

After rapid growth between the mid-1960s and the mid-1970s, the total number of bachelors degrees awarded each year remained essentially level until 1980, when growth resumed at a rate much slower than the earlier expansion. Over the same period, the annual production of bachelors degrees in the sciences and engineering grew more slowly than the overall trend.

Figure 10
Doctorate Degrees by Broad Area of Study (Thousands of Degrees)

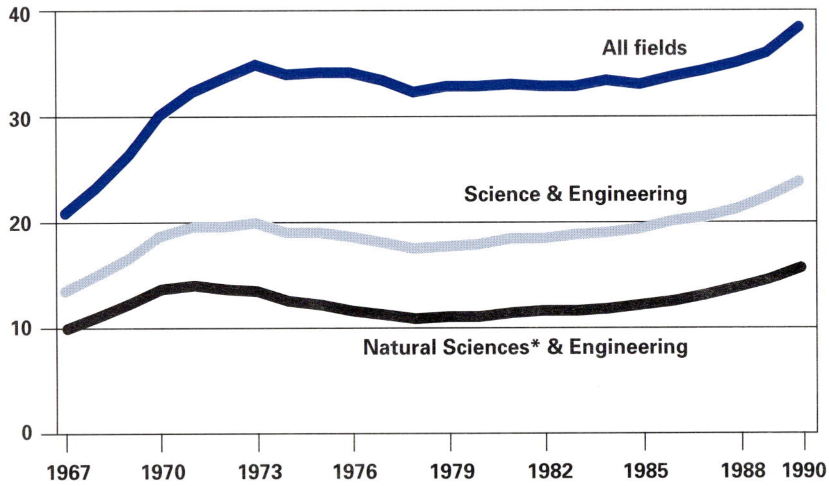

* Does not include medical and social sciences and psychology

Source: National Science Foundation and Department of Education

The total number of doctorates awarded each year exhibits a pattern similar to that for the bachelors degrees, albeit on a significantly smaller scale. However, the sciences and engineering account for a substantially larger fraction of the total doctorates than is the case for bachelors degrees and show a stronger upward trend in recent years.

Recommendation Five: Move People Between Industry and Universities

Accordingly, PCAST recommends that:

- universities and industry together, through a wide range of concerted actions, should exchange scientists and engineers at all levels — especially their very best — between the two sectors for substantial periods of time and repeatedly throughout their careers.

Specific activities and programs that contribute to that goal, and that should therefore be encouraged, include:

- undergraduate co-operative programs in industry;
- visiting professors from industry;
- dissertation research in industry laboratories;
- industry scientists mentoring graduate students; and
- faculty consulting for industry.

In sponsoring such activities, both industry and universities should seek the involvement of their most valued rather than their most expendable personnel. Both sectors will have to manage an increasing number of potential conflict-of-interest situations but, in doing so, should not stifle legitimate opportunities for greater interaction between university and industry personnel. In addition, any current or proposed federal or state regulation in this area should balance concerns about conflicts-of-interest against the value of greater university-industry interaction and cooperation.

VI. A Matter of the Best Scientific Talent: Tapping the Nation's Talent Base

> *Quite apart from the matter of fairness is the realization that we do not ever have enough bright, scientifically-minded young people to deprive any of them of a chance for a good education and viable career opportunities.*
>
> Barbara Webster, Professor, University of California - Davis
> July 1992 statement to PCAST

POTENTIALLY BRILLIANT YOUNG PEOPLE CAN BE FOUND THROUGHOUT THE POPULATION, WITHIN BOTH GENDERS AND EVERY RACE AND ETHNIC GROUP, IN EVERY ECONOMIC SITUATION, AND IN EVERY REGION OF THE COUNTRY.

Most important scientific discoveries were made by a small number of very gifted people who were provided the opportunity and time to pursue their intellectual interests. Potentially brilliant young people can be found throughout the population, within both genders and every race and ethnic group, in every economic situation, and in every region of the country. Stronger public policies must be designed to identify scientifically-gifted people at an early age and help them develop their talents, no matter what their circumstances.

Many factors influence the development of scientific and technological talent in the population. Most are beyond the scope of this report, but a few bear mentioning — high societal levels of maternal and child health, early learning programs, the availability of role models, competent science and mathematics teachers in the early grades, and the sympathetic portrayal of scientists and engineers in the media. Several of these factors, along with recommendations for federal action, have been dealt with in the recent PCAST report on "Learning to Meet the Science and Technology Challenge."

A range of local efforts, stimulated, for example, by a Department of Education program, are made to identify and help "gifted and talented" elementary and secondary school students. In addition, there are local and nationwide "science-fair" type competitions, national merit scholarships, and similar programs sponsored by industry, local or state government, or private foundations. While very worthwhile, these do not seem to be sufficient to identify, or provide support for, a high enough number of talented students to pursue careers in science and technology.

Japan and many European countries have a highly selective educational structure in which large proportions of students are systematically "weeded out" at relatively early ages and cannot easily proceed on to university-level work. The U.S. system has multiple reentry points where students whose interests and talents develop later than the norm, or whose circumstances change, can pursue their original or new educational goals. Too many promising students turn away from science at some early or middle stage of their education, only to find later that, despite the possibility of re-entry, "catching up" is quite difficult. Better ways must be found to reach out to these students and help them continue their educations.

Several types of federal and non-federal loan and grant programs are available for undergraduate support, but some students may not be able to qualify or may not be able, even with these programs, to afford to attend institutions of the caliber corresponding to their talents and interests. Portable undergraduate scholarships, by contrast, would allow a higher proportion of especially talented students to choose the schools that they wish to attend.

A student testifies before PCAST at one of six public meetings held nationwido.

At the graduate level, competitive, merit-based federal support for portable fellowships has enabled thousands of bright men and women to complete the doctoral degree and go on to become outstanding teachers and researchers in academia, industry, and government. Because they do not have to seek support as research assistants or teaching assistants, these fellowship holders may choose to attend — and are welcomed by — any university in the nation.

Changes to the federal tax code in 1986, and related actions by state and local authorities, have resulted in greatly increased taxation of scholarships, fellowships, and student stipends for research. Examples include payments for participating in hands-on undergraduate research sponsored by the National Science Foundation and summer participation at federal laboratories. Such taxation discourages students from accepting such support, complicates administration for the universities and the sponsoring authorities, and causes sponsors to raise payments — which reduces the number of persons they can sponsor.

If the United States is to continue to lead the world in basic scientific discoveries and in their exploitation, we will need to identify the most talented young people at the earliest possible time, encourage their interest in advanced education and science, and give them a sense of purpose as they pursue their education and career paths.

Recommendation Six: Identify and Nurture the Best Talent

Considering that potential science and engineering talent is distributed throughout the population, in diverse economic circumstances, and in every part of the country, PCAST recommends that:

- the federal government develop programs to award substantial numbers of portable undergraduate scholarships and graduate fellowships in science and engineering in each Congressional district. These awards would be made on a nationally competitive basis, using non-political, merit criteria, and would be designed to encourage greater numbers of outstanding students — throughout the nation — to pursue training, and then careers, in science and engineering.

As with the traditional program of National Science Foundation graduate fellowships, only citizens and permanent residents would be eligible; awardees would be able to attend any U.S. institution of their own choosing; and reasonable allowances would be included to cover institutional costs. The undergraduate program would include both beginning students and some who have completed one or two years of undergraduate work.

Moreover, PCAST notes that federal, state, and local tax policies should bolster, not undermine, the nation's investments — both public and private — in human capital. PCAST recommends that:

- federal, state, and local government end all taxation of scholarships, fellowships, and stipends for student participation in research.

Research-intensive universities can have a major effect on the development of scientific talent by educating inspiring teachers of precollege science and mathematics. PCAST, therefore, recommends that:

- research-intensive universities give greater emphasis to the education (including continuing education) of precollege teachers of science and mathematics; and

- the federal government provide scholarships or service-repayable loans to encourage talented students to attend research-intensive universities for careers as precollege teachers of science and mathematics.

Beyond the Horizon

This report is intended to address pressing problems that threaten the productive relationship between the federal government and research-intensive universities. We believe that the fundamental premises of this relationship are sound but that improvements are required for it to avoid deterioration and achieve its fullest potential.

We recognize the present time as one of tumultuous and profound changes in American society and in the world. The ending of the Cold War, the emergence of the European Community and nations of the Western Pacific Rim as economic powers, the changing demography of the American population, the ever increasing power of science and technology, and the growing awareness of new societal problems to which science and technology can be applied all require fresh and creative thinking of overall federal science and technology policy of which federal government-university relations are a part. While the current framework has served us well for four decades, it is far from obvious, as we move into the swifter current of the twenty-first century, that it will retain the validity it had when it was established in the middle of the twentieth century.

> **. . . IT IS FAR FROM OBVIOUS, AS WE MOVE INTO THE SWIFTER CURRENT OF THE TWENTY-FIRST CENTURY, THAT [THE CURRENT FRAMEWORK] WILL RETAIN THE VALIDITY IT HAD WHEN IT WAS ESTABLISHED IN THE MIDDLE OF THE TWENTIETH CENTURY.**

In our view, we must look beyond the immediate issues addressed in this report and conduct a broad national reexamination of the place of research and development in our national life including its fundamental rationale, goals, organization, funding, and administrative mechanisms. The importance of generating new knowledge and new technologies and of educating scientists, engineers, and the general public for the twenty-first century demands no less. It is the intention of PCAST, drawing fully on federal and state government as well as private sector expertise and experience, to undertake such a reexamination in a subsequent report.

APPENDIX A

SELECTED BIBLIOGRAPHY

AAU Ad Hoc Committee on Indirect Costs. *Indirect Costs Associated With Federal Support of Research on University Campuses: Some Suggestions for Change.* Washington, D.C.: Association of American Universities, December 1988.

Bok, Derek. "Reclaiming the Public Trust." *Change*, July/August 1992.

El-Khawas, Elaine. *Campus Trends, 1992.* American Council on Education: Washington, D.C., July 1992.

Government-University-Industry Research Roundtable. *The Academic Research Enterprise within the Industrialized Nations: Comparative Perspectives.* Washington, D.C.: National Academy Press, March 1990.

Government-University-Industry Research Roundtable. *Science and Technology in the Academic Enterprise: Status, Trends, and Issues.* Washington, D.C: National Academy Press, October 1989.

Government-University-Industry Research Roundtable. *Fateful Choices: The Future of the U.S. Academic Research Enterprise.* Washington, D.C.: National Academy Press, March 1992.

Government-University-Industry Research Roundtable. *Future National Research Policies Within the Industrialized Nations.* Washington, D.C: National Academy Press, April 1992.

Irvine, John; Martin, Ben; and Isard, Pheobe. *Investing in the Future.* Brookfield, Vermont: Edward Elgar Publishers, 1990.

Marshall, Eliot, and Palca, Joseph. "Cracks in the Ivory Tower." *Science*, Vol. 257, 28 August 1992.

National Academy of Sciences. *Responsible Science: Ensuring the Integrity of the Research Process.* Washington, D.C: National Academy Press, 1992.

National Science Board. *Science & Engineering Indicators.* Washington, D.C.: U.S. Government Printing Office, 1991.

National Science Board Commission on the Future of the National Science Foundation. *A Foundation for the 21st Century.* Washington, D.C., National Science Foundation, November, 1992.

National Science Foundation. *Vannevar Bush: Science — The Endless Frontier.* (Fortieth Anniversary Reprint). Washington, D.C., National Science Foundation, 1990.

National Science Foundation. *Scientific and Engineering Research Facilities at Universities and Colleges: 1992 (NSF 92-325).* Washington, D.C.: National Science Foundation, September 1992.

Raub, William F., "Cutting Red Tape on Research." *Issues in Science and Technology*, Vol. 5 No. 2 (Winter 1988-89).

Shapiro, Harold T. *Functions and Resources: The University of the Twenty-First Century.* Presentation at University of Chicago Symposium, October 1991.

Sowell, Thomas. "The Scandal of College Tuition." *Commentary*, August 1992.

U. S. House of Representatives, Committee on Science, Space and Technology. *Report of the Task Force on the Health of Research.* Washington, D.C.: U. S. Government Printing Office, July 1992.

U. S. House of Representatives, Select Committee on Children, Youth and Families. *College Education: Paying More and Getting Less.* (Press Release and Fact Sheet, September 14, 1992).

U. S. Office of Technology Assessment. *Federally Funded Research: Decisions for a Decade.* Washington, D.C.: Government Printing Office, May 1991.

White House Science Council Panel on the Health of U.S. Colleges and Universities. *A Renewed Partnership.* Washington, D.C.: Office of Science and Technology Policy, February 1986.

Appendix B

List of Meeting Hosts and Participants

We would particularly like to express our appreciation to the following persons who so generously hosted and helped to organize our public meetings, and presented their views on many of the topics discussed in the report.

Keith H. Brodie
President
Duke University

David H. Cohen
Provost
Northwestern University

William Cunningham
Chancellor, former President
The University of Texas at Austin

Frank Press
President
National Academy of Sciences

Chang Lin Tien
Chancellor
University of California at Berkeley

Charles Vest
President
Massachusetts Institute of Technology

We would also like to thank the following persons who presented their views at a public meeting, or who wrote to us.

David Adamay
President
Wayne State University

Chandra Adams
B.A. in Statistics
University of California at Berkeley

Daniel Albert
Professor Emeritus
University of Illinois at Urbana-Champaign

Alice Agogino
Assoc. Professor, Department of Mechanical Engineering
University of California at Berkeley

Jesus A. del Alamo
Professor, Department of Electrical Engineering
and Computer Science
Massachusetts Institute of Technology

Jonathan Allen
Director, Research Laboratory of Electronics
Massachusetts Institute of Technology

Jim Almond
Director, Center for High Performance Computing
University of Texas System

Greg Anderson
Associate Director for Systems and Planning
Massachusetts Institute of Technology

Bruce W. Arden
Dean
Engineering and Applied Science
University of Rochester

Kenneth I. Berns
American Society for Microbiology
Washington, DC

Stephen Bishop
Director, Microelectronics Laboratory
University of Illinois at Urbana-Champaign

Howard K. Birnbaum
Professor
Director, Materials Research Laboratory
University of Illinois at Urbana Champaign

Billy C. Black
President
Albany State College

S. Leslie Blatt
Clark University

Kristie A. Boering
Post Doctoral Fellow in Environmental Chemistry
Harvard University

Wolfgang-M. Boerner
Department of Electrical Engineering and Computer Science
The University of Illinois at Chicago

David D. Brown
Director, Department of Embryology
Carnegie Institution of Washington

Robert A. Brown
Professor
Department of Chemical Engineering
Massachusetts Institute of Technology

Peter Bruns
Director, Biological Sciences Division
Cornell University

John C. Buechner
Chancellor
University of Colorado at Denver

George Bugliarello
President
Polytechnic University

Bernard F. Burke
Professor
Massachusetts Institute of Technology

Robert L. Byer
Vice Provost/Dean of Research
Stanford University

George Campbell, Jr.
President
The National Action Council for Minorities in Engineering

Claude Canizares
Director, Center for Space Research
Massachusetts Institute of Technology

Don M. Carlton
President
Radian Corporation

G. Slade Cargill, III
President
Materials Research Society

Gail Cassell
American Society for Microbiology
Washington, DC

Alex Chen
Senior in Mathematics and Integrated Science
Northwestern University

Angie Ciaccia
Graduate Student in Pharmacology
University of North Carolina at Chapel Hill

Ralph Cicerone
Professor
Department of Geosciences
University of California at Irvine

James S. Clegg
Director
Bodega Marine Laboratory

Jonathan R. Cole
Provost
Columbia University

Mary Sue Coleman
Vice Chancellor for Graduate Studies and Research
University of North Carolina at Chapel Hill

Thomas C. Collins
Vice President
Oklahoma State University

Clare M. Cotton
Association of Independent Colleges and Universities
Massachusetts

John P. Crecine
President
Georgia Institute of Technology

Craig Crews
Graduate Student in Biochemistry
Harvard University

Alexander Crowell
Undergraduate Student in Physics and Mathematics
Wake Forest University

William H. Danforth
Chancellor
Washington University in St. Louis

Alexander V. d'Arbeloff
Chairman and President
Teradyne, Inc.
Dept. of Electrical & Computer Engineering
University of Illinois at Urbana-Champaign

Donald G. Davis, Jr.
Professor, Graduate School Library
Information Science
University of Texas at Austin

Stephen Doblin
College of Science & Technology
The University of Southern Mississippi

Earl Dowell
Engineering Deans Council
American Society for Engineering Education

Mary Maples Dunn
President
Smith College

James Economy
College of Engineering
University of Illinois at Urbana-Champaign

Robert H. Edwards
President
Bowdoin College

Kathleen Marie Eisenbeis
The University of Texas at Austin

Thomas Everhart
President
California Institute of Technology

Alan I. Faden
Dean
Georgetown University Medical Center

Gerald R. Fink
Director
Whitehead Institute

Carl Fisher
Graduate Student, Department of Mechanical Engineering
University of Texas at Austin

Markus Flik
Cryogenic Laboratory
Massachusetts Institute of Technology

David R. Ford
Vice President for Instructional Services and Dean of Faculty
Vincennes University

Joseph Froomkin
Economic Consultant
Chevy Chase, Maryland

Steve Fuller
Virginia Polytechnic Institute and State University

Robert W. Galvin
Chairman, Executive Committee
Motorola, Inc.

Norman E. Gaut
President and Chief Executive Officer
Picturetel Corporation

Robin M. Gavin, Jr.
President
Macalester College

Tajana George
Undergraduate Student in Electrical Engineering
North Carolina A & T State University

Melvin D. George
President
St. Olaf College

Martha W. Gilliland
Interim Vice President for Research
The University of Arizona

Sid Gilman
Professor and Chair, Dept. of Neurology
University of Michigan

Austin M. Gleeson
Moderator
Chairman and Professor, Department of Physics
University of Texas at Austin

Edwin L. Goldwasser
Acting Director
Computer-based Education Research Laboratory
University of Illinois at Urbana-Champaign

Hanna Holborn Gray
President
The University of Chicago

Henry Greenside
Associate
Department of Computer Science
Duke University

Grederick C. Greenwood
Director, Pacific Biomedical Research Center
University of Hawaii at Manoa

Phillip A. Griffith
Director
Institute for Advanced Study

Alan E. Guskin
President
Antioch University

David H. Guston, Ph.D. candidate
Massachusetts Institute of Technology

Norman Hackerman
Chairman
Texas Higher Education Coordinating Board
Committee on Research Programs

Charles Hamner
President
North Carolina Biotech Center

Linda Hansen
B.A. in Chemistry
Stanford University

Robin Hanson
NASA Ames Research Center

Kerry Haynie
Graduate Student in Political Science
University of North Carolina at Chapel Hill

Anthony E. Hechanova
President
American Nuclear Society Student Chapter
Massachusetts Institute of Technology

Karl Hess
Director, Center for Computational Electronics
University of Illinois at Urbana-Champaign

Carl E. Hewitt
Professor
Massachusetts Institute of Technology

Jacqueline N. Hewitt
Professor
Department of Physics
Massachusetts Institute of Technology

Gary J. Hill
Research Associate
Department of Astronomy
University of Texas at Austin

Larry Hochhaus
Professor
Oklahoma State University

Carol Hollenshead
Director
Center for the Education of Women
The University of Michigan

Barbara D. Holmes
President
Milwaukee Area Technical College

Hal Hopfenberg
Executive Assistant to Chancellor
North Carolina State University

Rustin Howard
President, Phyton Catalytic

Robert Howe
Professor
Division of Applied Sciences,
Harvard University

Judy Hoyt
Research Associate, Electrical Engineering
Stanford University

Mark V. Hurwitz
Assistant Research Astronomer
University of California at Berkeley

John S. Hutchinson
Professor
Department of Chemistry
Rice University

William P. Hytche
President
University of Maryland Eastern Shore

Ettore F. Infante
Provost
University of Minnesota
John Ingraham
American Society for Microbiology

John Jacobson
President
Hope College

Franklyn D. Jenifer
President
Howard University

W. Kenneth Jenkins
Director, Coordinated Science Lab
University of Illinois at Urbana-Campaign

F. Scott Johnson
Graduate Student in Materials Science
North Carolina State University

Todd Johnson
Senior in Industrial Engineering
Northwestern University

Patricia Culver Keane
Graduate Student in Chemistry
Northwestern University

Elizabeth T. Kennan
President
Mt. Holyoke College

Nannerl O. Keohane
President
Wellesley College

C. William Kern
Vice President for Research and Dean of
Graduate School
Northwestern University

Joe L. Key
Vice President for Research
The University of Georgia

George D. Klein
Professor
University of Illinois at Urbana-Champaign

Rebecca Richards Kortum
Assistant Professor, Department of Electrical
and Computer Engineering
University of Texas at Austin

Daniel Koshland
Professor Emeritus
Department of Molecular & Cell Biology
University of California at Berkeley

John W. Kuykendall
President
Davidson College

Gerald Lame
Research Associate, Linguistics Department
University of Texas

Donald N. Langenberg
Chancellor
University of Maryland System

Jules LaPidus
President
Council of Graduate Schools

James Leheny
Associate Chancellor
University of Massachusetts at Amherst

Morris W. Leighton
Chief
Illinois State Geological Survey

Peter C. Magrath
President
National Association of State Universities and Land Grant Colleges

Thomas J. Malone
President
MILLIKEN

Thomas F. Malone
Director
Sigma Xi Center

G. Ali Mansoori
Professor
The University of Illinois at Chicago

Hans Mark
former Chancellor
The University of Texas System

David McClay
Professor
Department of Zoology
Duke University

David McNamara
Chairman, Science Department
Shaker Heights High School, Cleveland, Ohio

Linda McPheron
Department of Entomology
University of California at Berkeley

Mary Patterson McPherson
President
Bryn Mawr College

David F. Mears
Director
University of California

Chad A. Mirkin
Assistant Professor, Chemistry
Northwestern University

W. H. Mobley
President
Texas A. & M. University

Bradley Moore

Calvin Moore

William V. Muse
President
Auburn University

Jeanne Narum
Director
The Independent Colleges Office

Sean Xavier Neath
Graduate Student in Biochemistry and Nutrition
University of Chicago

Darin Nelson
Graduate Student in Neurobiology
Duke University

Gordon Nelson
Chairman
Council of Scientific Society Presidents

Dava Newman
Recent Ph.D. recipient in Aeronautics and Astronautics
Massachusetts Institute of Technology

Sarah Winans Newman
Associate Vice President for Research
University of Michigan

James Niedel
Senior Vice President and Director
Glaxo Research Institute

Janet Osteryoung
Head of Department of Chemistry
North Carolina State University

F. Michael Pestorias
Director, Applied Research Laboratories
University of Texas at Austin

Robert G. Petersdorf
President
Association of American Medical Colleges

David Pierce
President
American Association of Community and Junior Colleges

David Pramer
American Society for Microbiology

P. Buford Price
Dean, Physical Sciences
University of California at Berkeley

Dale Purves
Chairman
Department of Neurobiology
Duke University

Charles E. Putman
Professor of Medicine
Duke University

A. Kenneth Pye
President
Southern Methodist University

Ralph Quatrano
Chairman
Department of Biology
University of North Carolina

Theda M. Daniels Race
Electrical Engineering
Duke University

Gina Raimondo
Undergraduate in Economics
Harvard University

Frank H. T. Rhodes
President
Cornell University

Herbert Richardson
Chancellor
Texas A. & M. University System

John Richmond

Robert M. Rosenzweig
President
Association of American Universities

David Salzman
Harvard University

Patrick Scannon
Founder & Vice Chairman for Science and Medical Affairs
XOMA Corporation

Margo Seltzer
Assistant Professor
Harvard University

Charles Shank
Director
Lawrence Berkeley Laboratory

W. A. Sibley
Vice President
The University of Alabama at Birmingham

Robert Simoni
Professor, Department of Biological Sciences
Stanford University

William J. Spencer
President and Chief Executive Officer
Sematech

Ellen Spertus
Graduate Student in the Laboratory for Computer Science
Massachusetts Institute of Technology

Allen P. Spelte
President
The Council of Independent Colleges

Peter W. Stanley
President
Pomona College

Harvey J. Stapleton
Interim Vice Chancellor for Research
University of Illinois at Urbana-Champaign

E. R. Stout
Associate Provost for Research
Virginia Polytechnic Institute and State University

Rudi Strickler
Distinguished Professor, Biological Sciences
University of Wisconsin - Milwaukee

Jim Sullivan
Massachusetts Institute of Technology

Patricia B. Swan
Vice Provost
Iowa State University

Michael Tanner
Academic Vice Chancellor
University of California at Santa Cruz

Mike Thomas
Georgia Tech University

Joan M. Torykian
Armenian Women's Archives
Berkeley, CA

Daniel E. Tosteson
Dean
Harvard Medical School

Timothy N. Trick
Head
Department of Electrical & Computer Engineering
University of Illinois at Urbana-Champaign

Charles M. Vest
President
Massachusetts Institute of Technology

Joe B. Wyatt
Chancellor
Vanderbilt University

Graham Walker
Professor
Department of Biology
Massachusetts Institute of Technology

Paul Wallace
Department of Geology & Geophysics
University of California at Berkeley

Barbara Webster
Department of Agronomy & Range Science
University of California, Davis

Steven Weinberg
Professor
Department of Physics
University of Texas at Austin

Derek H. Willard
Interim Vice President
The University of Iowa

Virginia Wilson
Undergraduate Student, Department of Chemistry and Biochemistry
University of Texas at Austin

Susanne Woods
Vice President and
Dean of the College, Franklin and Marshall College

Jacqueline D. Woolley
Assistant Professor
Department of Psychology
University of Texas at Austin

Mark S. Wrighton
Provost
Massachusetts of Institute of Technology

Karen M. Yarbrough
Vice President for Research and Planning
The University of Southern Mississippi

Annette M. Yonke
Associate Professor
College of Medicine at Chicago

Elizabeth A. Zinser
President
The University of Idaho

Nathan Zook
University Fellow - Mathematics
University of Texas at Austin

Aaron Zorn
Graduate Student, Department of Zoology
University of Texas at Austin

ISBN 0-16-038257-2

90000

9 780160 382574